GENERAL RELATIVITY
A FIRST EXAMINATION

Other Related Titles from World Scientific

*Interacting Gravitational, Electromagnetic, Neutrino and Other Waves:
In the Context of Einstein's General Theory of Relativity*
by Anzhong Wang
ISBN: 978-981-121-148-5

Origin and Evolution of the Universe: From Big Bang to ExoBiology
Second Edition
edited by Matthew A. Malkan and Ben Zuckerman
ISBN: 978-981-120-645-0
ISBN: 978-981-120-772-3 (pbk)

Stars and Space with MATLAB Apps: With Companion Media Pack
by Dan Green
ISBN: 978-981-121-602-2
ISBN: 978-981-121-635-0 (pbk)

The Dark Energy Survey: The Story of a Cosmological Experiment
edited by Ofer Lahav, Lucy Calder, Julian Mayers and Josh Frieman
ISBN: 978-1-78634-835-7

Loop Quantum Gravity for Everyone
by Rodolfo Gambini and Jorge Pullin
ISBN: 978-981-121-195-9

Second Edition

GENERAL RELATIVITY
A FIRST EXAMINATION

Marvin Blecher

Virginia Tech, USA

World Scientific

NEW JERSEY · LONDON · SINGAPORE · BEIJING · SHANGHAI · HONG KONG · TAIPEI · CHENNAI · TOKYO

Published by

World Scientific Publishing Co. Pte. Ltd.

5 Toh Tuck Link, Singapore 596224

USA office: 27 Warren Street, Suite 401-402, Hackensack, NJ 07601

UK office: 57 Shelton Street, Covent Garden, London WC2H 9HE

Library of Congress Cataloging-in-Publication Data
Names: Blecher, M. (Marvin), 1940–   author.
Title: General relativity : a first examination / Marvin Blecher, Virginia Tech, USA.
Description: Second edition. | New Jersey : World Scientific Publishing Co. Pte. Ltd., [2021] |
    Includes bibliographical references and index.
Identifiers: LCCN 2020019202 | ISBN 9789811220432 (hardcover) |
    ISBN 9789811221088 (paperback) | ISBN 9789811220449 (ebook) |
    ISBN 9789811220456 (ebook other)
Subjects: LCSH: General relativity (Physics)--Textbooks.
Classification: LCC QC173.6 .B56 2021 | DDC 530.11--dc23
LC record available at https://lccn.loc.gov/2020019202

British Library Cataloguing-in-Publication Data
A catalogue record for this book is available from the British Library.

For any available supplementary material, please visit
https://www.worldscientific.com/worldscibooks/10.1142/11831#t=suppl

Desk Editor: Ng Kah Fee

Typeset by Stallion Press
Email: enquiries@stallionpress.com

Printed in Singapore

I dedicate this work to Freda, the love of my life, the fount of my happiness and contentment, and to my family whose love sustains me.

Preface

Upon retirement, I sought a learning project. My career was spent as an experimental physicist studying nuclei and particles. Throughout, I was deeply impressed by the connection between my area and General Relativity and cosmology. Now, I had the time to dig into the details of those subjects. When I asked, the Virginia Tech Physics Department granted my request to teach the introductory semester course in General Relativity. I did so three times and found that nothing reduces knowledge deficiency like teaching smart students.

Many excellent texts were available for my studies. However, there was too much material in them for students to cover in a single semester. So I developed my own set of notes, that explained the essentials of the subject in the course time period. This text evolved from those notes. I'm grateful to my students because their questions pushed me to explain difficult concepts as transparently as possible. I'm also grateful to many of my colleagues who participated in discussions, helped with proofs and critiqued some of my material.

Einstein's theory is now a century in age. It is well tested, but still inspires significant theoretical and experimental work. And it definitely intrigues students. Advanced undergraduate physics majors, first year physics graduate and engineering students have taken the course. Physics Department faculty and faculty from other departments have sat in on many lectures. I thank them all for their comments and questions. They helped me acquire a deeper understanding of the subject.

Much of the material in this book is found in various forms in other texts. However, this text contains much that is novel, and many more steps than usual are included in proofs.

In Chapter 2, the twin problem with acceleration is worked out by the twin at rest. Later in Chapter 5, the accelerating twin uses the full power of General Relativity to predict the same time ratio result.

The way gravity affects time is first discussed for weak gravity via conservation of energy using a Newtonian formulation with relativistic mass. In Chapter 5, weak gravity in General Relativity is discussed and the way gravity affects time is rigorously covered. The Schwarzschild metric is obtained and in Problem 5 of that chapter, students are asked to solve the Schwarzschild problem with the cosmological constant included. Then they are asked to show that in weak gravity a very small repulsive Newtonian force arises.

In Chapter 7, gravitational waves are discussed. Just as the first edition to this book was going to press, the LIGO experiment announced the first direct detection of such a wave. Luckily, I was able to include this finding. The theory behind the Nobel prize winning, gravitational wave indirect detection results, from the Hulse–Taylor binary pulsar, is worked out with elliptical orbits. Other texts that discuss this experiment have used circular orbits. As the eccentricity is large, the latter orbits disagree with experiment.

In Chapter 8 on black holes and Kerr space, an example, based on the film "Interstellar", is presented. The example discusses why a large gravitational time dilation is possible near a spinning, but not a static black hole. Geodesics in Kerr space are also worked out.

In Chapter 9 on cosmology, the results of numerical integrations using the current data for all the energy densities are discussed. This allows the students to peer into both the past and the future for values of the universal scale factor, the Hubble parameter, the age of the universe and the horizon distance.

Acknowledgments

I'm grateful to colleagues N. Arav, P. Huber, J. Link, and D. Roper, who participated in discussions with me, and read a chapter. I especially thank Tatsu Takeuchi who helped with proofs, read, and critiqued many of my chapters; and John Simonetti who cleared up many concepts, and explained details of the astronomical experiments discussed in the text. Eric Sharpe and James Gray welcomed many questions, and read a chapter. I'm indebted to Ms Samantha Spytek, an undergraduate physics major at Virginia Tech, and to Robin Blecher, for preparing many of the figures.

Contents

3.   Covariant Differentiation, Equations of Motion   41

4.   Curvature   55

5.   Gravity and General Relativity   69

6.   Classic Solar System Tests of General Relativity   91

7.   Gravitational Waves   107

Chapter 1

Review of Special Relativity

1.1 Introduction

The theory of Special Relativity (SR) was introduced by A. Einstein in 1905. It deals with the observations of inertial observers in the absence of gravity. The theory of General Relativity (GR) that includes gravitation, and thus acceleration, was published in 1915. For English translations, see Einstein (1905). The latter theory predicted the deflection of light near a massive body, like the sun. Shortly after the end of the first world war, a British team, led by A. S. Eddington, confirmed this startling prediction. This made Einstein world famous, even among people who had no particular interest in science.

In relativity, an observation is the assignment of coordinates x^μ, $\mu = 0, 1, 2, 3$, for the time and space location of an event. Space is continuous, and functions of the coordinates can be differentiated. Upon partial differentiation with respect to one of the coordinates, the others are held constant. This insures that the coordinates are independent,

$$x^\mu{}_{,\nu} \equiv \frac{\partial x^\mu}{\partial x^\nu} = \delta^\mu{}_\nu = \delta_\nu{}^\mu = 1, \quad \mu = \nu, \quad \delta^\mu{}_\nu = 0, \quad \mu \neq \nu. \quad (1.1)$$

As will be seen $\delta^\mu{}_\nu$ is the Kronecker delta tensor. The superscript, subscript indexes are termed contravariant, covariant. Note the shorthand notation for the partial derivative, by use of a comma. Such a shorthand will keep some of the formulas of GR, with many partial derivatives, to a reasonable length.

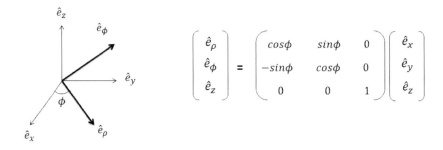

$$\begin{pmatrix} \hat{e}_\rho \\ \hat{e}_\phi \\ \hat{e}_z \end{pmatrix} = \begin{pmatrix} cos\phi & sin\phi & 0 \\ -sin\phi & cos\phi & 0 \\ 0 & 0 & 1 \end{pmatrix} \begin{pmatrix} \hat{e}_x \\ \hat{e}_y \\ \hat{e}_z \end{pmatrix}$$

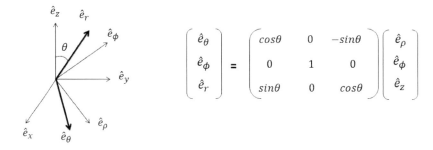

$$\begin{pmatrix} \hat{e}_\theta \\ \hat{e}_\phi \\ \hat{e}_r \end{pmatrix} = \begin{pmatrix} cos\theta & 0 & -sin\theta \\ 0 & 1 & 0 \\ sin\theta & 0 & cos\theta \end{pmatrix} \begin{pmatrix} \hat{e}_\rho \\ \hat{e}_\phi \\ \hat{e}_z \end{pmatrix}$$

Fig. 1.1   Rotation relations for changing unit vectors from one coordinate system to another.

In rectangular coordinates $x^\mu = (t, x, y, z)$. In relativity, one may transform to other coordinate systems, or to the rectangular coordinates of other reference frames. These coordinates will be labeled by primes, $x^{\mu'}$. Curvilinear coordinates are particularly useful, and a rotation carries you from one set of coordinates to the other. In cylindrical coordinates $x^{\mu'} = (t, \rho, \phi, z)$ because as illustrated in Fig. 1.1, the rotation changes the direction indicating unit vectors $(\hat{e}_x, \hat{e}_y) \rightarrow (\hat{e}_\rho, \hat{e}_\phi)$. Similarly, for spherical coordinates $x^{\mu''} = (t, \theta, \phi, r)$, $(\hat{e}_\rho, \hat{e}_z) \rightarrow (\hat{e}_\theta, \hat{e}_r)$. Other texts employ an extra renaming, and take $x^{\mu''=0-3} = t, r, \theta, \phi$, but a rose by any name would smell sweet. The reader can prove that the components of 3-vectors, when written in terms of unit vectors $\vec{V} = V^x \hat{e}_x + V^y \hat{e}_y + V^z \hat{e}_z$, are transformed by rotations in the same way as the unit vectors.

The time of the event is read on a clock at rest with respect to the observer, at the spatial coordinates of the event. In the inertial frames of

SR, an observer may suppose that there are synchronized clocks at rest at every point in space. This would not be the case when gravity is taken into account, since clocks run at different rates, in a region of varying gravitational strength. Simultaneous events for a given observer are those occurring at the same time on the clocks nearest them, that are at rest with respect to the observer.

As will be seen, we live in spacetime of four dimensions where space and time mix with each other. Thus, spacetime vectors have four components. By analogy with three space, take as the rectangular components of the contravariant position vector r^μ the coordinates x^μ. Then dx^μ must also be the components of the displacement vector dr^μ, as the difference of two vectors is also a vector. As of now, these are the only vectors known to us.

In later chapters, when gravity or acceleration may be acting, the symbol for a set of coordinates may have a bar $x^{\bar\mu}$. This will indicate that these coordinates are those of an inertial observer using rectangular coordinates. This is the case for an SR reference frame that is arbitrarily large. When gravity is present the bar will usually be absent. However, as will be shown, even when gravity is present, at any point in space one can find a locally inertial frame. That frame may need be arbitrarily small. The rectangular coordinates of that frame will have the bar when the set of coordinates is described. Other coordinate systems or rectangular coordinates of other inertial reference frames will be represented by the bar and one or more primes $x^{\bar\mu'}$. In such cases, the components of vectors $r^{\bar\mu}$ and higher order tensors $T^{\bar\mu\bar\nu}$ will also have bars.

Einstein developed SR from two postulates: (1) the laws of physics are the same for all inertial observers no matter their relative velocities; (2) all inertial observers measure the same speed of light in vacuum $c = 3 \times 10^8$ m/s. It is the second postulate that causes conflict with the Newtonian concept of time flowing independent of everything else. This leads to the observation, that events simultaneous to one observer may not be so to another. Also c becomes the limiting speed in order to preserve causality. In GR, the word "inertial" is removed, and the principle of equivalence: that no gravitational effect is experienced when freely falling in a region of uniform gravitational strength, must be taken into account.

In hindsight, it is easy to see where the postulates come from. Various inertial observers, in empty space and in relative motion, perform electromagnetic experiments in their own rest frames. They find that the equations of Maxwell for the electric, magnetic fields $(\vec{E}, \vec{B})$ explain the results.

In vacuum, they use the empty space permittivity, permeability (ϵ_0, μ_0). In MKS units, each finds that they lead to a wave equation, with a unique velocity,

$$\vec{\nabla} \cdot \vec{E} = 0 = \vec{\nabla} \cdot \vec{B}, \quad \vec{\nabla} \times \vec{E} = -\frac{\partial \vec{B}}{\partial t}, \quad \vec{\nabla} \times \vec{B} = \mu_0 \epsilon_0 \frac{\partial \vec{E}}{\partial t},$$

$$0 = \nabla^2 \vec{B} - \mu_0 \epsilon_0 \frac{\partial^2 \vec{B}}{\partial t^2}, \quad v = (\mu_0 \epsilon_0)^{-1/2} = c.$$

As c is so special in SR and GR, it is convenient to work in a system of units where velocities are dimensionless and $c = 1$. Then time is expressed in meters (m), like the other coordinates, and acceleration is expressed in inverse meters:

$$c = 1 = 3 \times 10^8 \, \mathrm{m\,s}^{-1},$$

$$s = 3 \times 10^8 \, \mathrm{m}, \tag{1.2}$$

$$a = \mathrm{m\,s}^{-2} = 0.111 \times 10^{-16} \, \mathrm{m}^{-1}.$$

Similarly in GR, Newton's gravitational constant G is so special, that it is convenient to also use $G = 1$. This leads to the natural system of units. Here other mechanical quantities, like mass, energy, momentum, and angular momentum, can be expressed in meters to the correct power:

$$1 = \frac{G}{c^2} = \frac{6.674 \times 10^{-11} \, \mathrm{m^3 \, kg}^{-1} \, \mathrm{s}^{-2}}{[3 \times 10^8]^2 \, \mathrm{m^2 \, s}^{-2}}$$

$$= 0.742 \times 10^{-27} \, \mathrm{kg}^{-1} \, \mathrm{m},$$

$$M = \mathrm{kg} = 0.742 \times 10^{-27} \, \mathrm{m}, \tag{1.3}$$

$$E = \mathrm{kg \, m^2 \, s}^{-2} = 0.824 \times 10^{-44} \, \mathrm{m}.$$

Suppose a result is obtained in naturalized units for $\hbar = h/(2\pi) = 2.612 \times 10^{-70} \, \mathrm{m}^2$, where h is Planck's constant. One can calculate the value in MKS units by noting that in this system the units are those of angular momentum $\mathrm{kg \, m^2 \, s}^{-1}$. Multiply the value in natural units by unity, with a quantity that expressed in MKS units, will give the desired units,

$$\hbar = 2.612 \times 10^{-70} \, \mathrm{m}^2 [c/(G/c^2)]$$

$$= \frac{(2.612 \times 10^{-70})(3 \times 10^8)}{0.742 \times 10^{-27}} \, \mathrm{m}^2 (\mathrm{m/s})/(\mathrm{m/kg})$$

$$= 1.056 \times 10^{-34} \, \mathrm{kg \, m^2 \, s}^{-1}.$$

1.2 Lorentz Transform

Two observers O and O′ are considered. They use parallel axes and rectangular coordinates. Rotations, like those in Fig. 1.1, allow them to align their z-axes along the relative velocity. O uses x^μ, and says O′ is moving in the z-direction with speed $V(<1)$, while O′ uses $x^{\mu'}$, and says O is moving in the $-z$-direction with speed V.

When their origins overlapped, the clocks were synchronized $t = x^0 = t' = x^{0'} = 0$. In this geometry, $(x, y) = (x', y')$ or $x^{1,2} = x^{1',2'}$, as there is no relative motion in these directions. However, $c = 1$ for both observers, so space and time are interconnected, and now termed spacetime. If O′ says that events led to changes in coordinates $dz' = dx^{3'}$ and $dt' = dx^{0'}$, the components of the displacement vector $dr^{\mu'}$, then O would calculate from the chain rule of differential calculus,

$$dx^3 = dz = \frac{\partial z}{\partial z'}dz' + \frac{\partial z}{\partial t'}dt' + \frac{\partial z}{\partial x'}dx' + \frac{\partial z}{\partial y'}dy' \equiv x^3{}_{,\mu'}\,dx^{\mu'}$$

$$= x^3{}_{,3'}\,dx^{3'} + x^3{}_{,0'}\,dx^{0'}, \tag{1.4}$$

$$dx^0 = dt = \frac{\partial t}{\partial z'}dz' + \frac{\partial t}{\partial t'}dt' + \frac{\partial t}{\partial x'}dx' + \frac{\partial t}{\partial y'}dy' \equiv x^0{}_{,\mu'}\,dx^{\mu'}$$

$$= x^0{}_{,3'}\,dx^{3'} + x^0{}_{,0'}\,dx^{0'}. \tag{1.5}$$

One notes that, similar to rotations, this transform can be represented by matrix multiplication,

$$\begin{pmatrix} dx^0 \\ dx^1 \\ dx^2 \\ dx^3 \end{pmatrix} = \begin{pmatrix} x^0{}_{,0'} & 0 & 0 & x^0{}_{,3'} \\ 0 & 1 & 0 & 0 \\ 0 & 0 & 1 & 0 \\ x^3{}_{,0'} & 0 & 0 & x^3{}_{,3'} \end{pmatrix} \begin{pmatrix} dx^{0'} \\ dx^{1'} \\ dx^{2'} \\ dx^{3'} \end{pmatrix}.$$

This is a linear transform. The vector components appear to the power unity. The coefficients, the partial derivatives multiplying the O′ vector components, are relations between the coordinates of the different frames. They are independent of the vectors. Such a rule for vector transformation is not limited to rotations. It works for any coordinate transformation, and any vector. If each vector had a different transformation rule, there would be no theory. If a set of four quantities V^μ do not transform as above, then they are not components of a vector.

This transformation does not yield vectors, written in terms of unit vectors $\hat{e}_\mu$, but in terms of basis vectors $\vec{e}_\mu$. The relationship between basis and unit vectors, for rotations, is explored in Problem 3. There, the components of the 3-vector $d\vec{r}$ in rectangular, cylindrical, and spherical coordinates, in terms of unit and basis vectors, lead to the relationship. In relativity, basis vector components are used.

Note the summation over an index definition, in Eqs. (1.4) and (1.5), requires the same index repeated as both contravariant and covariant. In the case of the partial derivative, a covariant index results from the contravariant index in the denominator. Coordinates, by tradition, are always written with contravariant indexes as opposed to tensors, vectors are tensors of rank 1, that have both types of indexes.

Let both observers O and O$'$ concentrate on a light ray,

$$
1 = \left(\frac{dx^1}{dx^0}\right)^2 + \left(\frac{dx^2}{dx^0}\right)^2 + \left(\frac{dx^3}{dx^0}\right)^2
$$

$$
= \left(\frac{dx^{1'}}{dx^{0'}}\right)^2 + \left(\frac{dx^{2'}}{dx^{0'}}\right)^2 + \left(\frac{dx^{3'}}{dx^{0'}}\right)^2 ,
$$

$$
(d\tau)^2 \equiv (dx^0)^2 - (dx^1)^2 - (dx^2)^2 - (dx^3)^2 = 0, \tag{1.6}
$$

$$
(d\tau')^2 \equiv (dx^{0'})^2 - (dx^{1'})^2 - (dx^{2'})^2 - (dx^{3'})^2 = 0. \tag{1.7}
$$

Thus,

$$
(d\tau)^2 = (d\tau')^2,
$$
$$
(dx^0)^2 - (dx^3)^2 = (dx^{0'})^2 - (dx^{3'})^2. \tag{1.8}
$$

Applying Eqs. (1.4) and (1.5) to the above equation,

$$
(dx^0)^2 = (x^0{}_{,0'}\, dx^{0'})^2 + (x^0{}_{,3'}\, dx^{3'})^2 + 2x^0{}_{,0'}\, x^0{}_{,3'}\, dx^{0'}\, dx^{3'},
$$

$$
(dx^3)^2 = (x^3{}_{,0'}\, dx^{0'})^2 + (x^3{}_{,3'}\, dx^{3'})^2 + 2x^3{}_{,0'}\, x^3{}_{,3'}\, dx^{0'}\, dx^{3'},
$$

$$
1 = (x^0{}_{,0'})^2 - (x^3{}_{,0'})^2 \equiv \cosh^2 \alpha - \sinh^2 \alpha
$$

$$
= -(x^0{}_{,3'})^2 + (x^3{}_{,3'})^2 \equiv \cosh^2 \beta - \sinh^2 \beta,
$$

$$
0 = \cosh \alpha \sinh \beta - \sinh \alpha \cosh \beta = \sinh(\beta - \alpha) \rightarrow \alpha = \beta.
$$

The above results force Eq. (1.8) to also hold, for other than light travel. So $d\tau$ is an invariant, a tensor of rank 0, numerically the same in all frames. The invariant $d\tau$ is the proper time, that read on a clock at rest with respect to the observer. For light travel, assumed in vacuum unless otherwise noted, $d\tau = 0$. So photons never run out of time, and some created in the very early universe, are still around. The interval $(dS)^2 \equiv -(d\tau)^2$. Although τ is a member of the Greek alphabet, and could be used as an index, it is reserved for the proper time.

In order to calculate $\sinh \alpha$ and $\cosh \alpha$, O concentrates on the position of O'. In time period dt, O' changes position $dz = V dt$. However, O' says, "I am at rest while my clock has advanced by dt'." Using Eqs. (1.4) and (1.5),

$$dt = dt' \cosh \alpha, \quad V dt = dt' \sinh \alpha,$$

$$V = \tanh \alpha,$$

$$\gamma \equiv \cosh \alpha = (1 - V^2)^{-1/2} = x^0{}_{,0'} = x^3{}_{,3'}, \tag{1.9}$$

$$\gamma V = \sinh \alpha = x^0{}_{,3'} = x^3{}_{,0'}. \tag{1.10}$$

The reverse transform from unprimed to primed coordinates just requires,

$$t \leftrightarrow t', \ z \leftrightarrow z', \ V \to -V,$$

$$\gamma = x^{0'}{}_{,0} = x^{3'}{}_{,3}, \ -\gamma V = x^{0'}{}_{,3} = x^{3'}{}_{,0}. \tag{1.11}$$

Note that for low speeds $(1 - V^2)^{-1/2} \to 1$, and the Galilean transform is recovered.

1.3 Physics Consequences

First, consider simultaneity. O and O' are coincident. O' says two events occur at the same time so $dt' = 0$, but at positions from the origin $\pm dz'$. O says that the time differences from zero of the two events are,

$$dt_\pm = \pm \gamma V dz', \quad dt_+ - dt_- \neq 0.$$

In general, observers in relative motion do not agree on simultaneity. Only if events are spatially coincident will observers so agree. Thus, to compare times, clocks at the same spatial position have to be compared. Also note, the present position of O', when all clocks are synchronized, is connected with all points in the past, present, and future of O. In spacetime, all space and time points are available. Time isn't a special quantity, it's just a one-dimensional projection of spacetime.

Next consider causality. In the unprimed frame, observer O fires a bullet at $t = 0$ from the origin. It hits a target at time dt and position $dz = v_b dt$, where v_b is the speed of the bullet. O' also says the bullet was fired from the origin at $t' = 0$, as that's where the clocks were synchronized. O' says the target was hit at

$$dt' = \gamma(dt - V[v_b dt]) = dt\gamma(1 - Vv_b).$$

According to O', if $dt' < 0$, the target is hit before the bullet is fired. That would violate causality, and can happen only if $v_b > 1$. Thus, $c = 1$ is the limiting speed.

Now consider the comparison of clocks. Observer O and clock A are spatially coincident. Other clocks B, C, ... are at rest with respect to O, and read the same time as A. Observer O' and clock B' are spatially coincident. Other clocks A', C', ... are at rest with respect to O', and read the same time as B'. All the clocks are synchronized, when A and B' are spatially coincident. Since O says A is at rest, when it has ticked off a time period $dt = d\tau$, as in Fig. 1.2, what will the clocks of O' read? O concludes that clock C' is now spatially coincident with clock A. The times on these two clocks can be compared. O says A has ticked off $d\tau$, while remaining

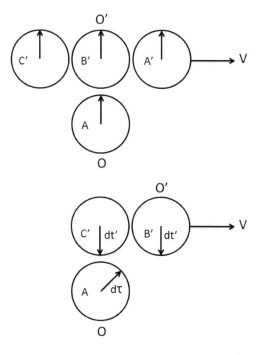

Fig. 1.2   Comparison of clocks. The view of O.

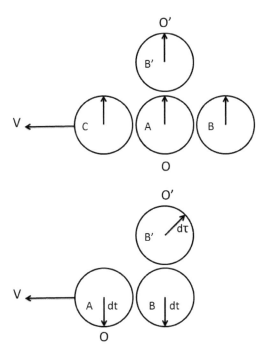

Fig. 1.3 Comparison of clocks. The view of O'.

at rest. From the Lorentz transform, the time ticked off on C' is

$$dt' = \gamma d\tau > d\tau.$$

This result is called time dilation. Time and all processes related to it run slower for O, who considers herself at rest, as compared with those of O', who O sees moving. So A ticks slower. Also the heart rate, and other biological rates of O are slower.

Alternately, O' considers B' at rest, as in Fig. 1.3. When it has ticked off time $dt' = d\tau$, what will the clocks at rest with respect to O read? According to O', clock B has moved into spatial coincidence with clock B'. The times on these two clocks can be compared. From the Lorentz transform,

$$dt = \gamma d\tau > d\tau.$$

Now time and all processes related to it run slower for O', who considers himself at rest, as compared with those of O, who O' sees moving. So B' ticks slower, and the biological rates of O' are slower.

Strange as it seems, both observers are correct about these particular measurements. There is no third observer who could decide the case in favor of O or O'.

Time dilation has been experimentally verified in the laboratory with particles called muons. Muons at rest have a mean lifetime of about 660 m. Thus, if you begin with a large number of muons at rest N_0, the number N left after time t m is $N/N_0 = \exp(-t/660)$. Then, the half life for muons $t_{1/2} = 660 \ln 2$ m is the time when half of the starting muons remain. This result must be interpreted as the probability of muon survival. After a time $t = nt_{1/2}$ experimenters expect $N/N_0 = (1/2)^n$.

In the laboratory, muons can be created with speed $V \approx 1$, and almost all are observed to travel far longer distances than 660 m. This means that the clock attached to the muon (B') has not yet ticked off even one mean lifetime. The clocks (A, B, C) ... attached to the laboratory, that the muon sees moving, have ticked off a much longer time period.

One feature of SR that hasn't been directly confirmed is length contraction. Suppose a rod of length dz' is at rest in O'. That length could be measured by O', by measuring the coordinate of each end at arbitrary times. However, O sees the rod moving, and so must measure the coordinates of the two ends at the same time $dt = 0$. This determines the moving length dz,

$$dz' = \gamma dz,$$

$$L(\text{rest}) = \gamma L(\text{moving}) > L(\text{moving}). \tag{1.12}$$

By considering the muons, confirmation of this result can be inferred. In the laboratory, the speedy muons easily travel much more than 660 m. This is the distance between points i and j in Fig. 1.4. However, according to the muons, these points are moving, and the distance between them is length contracted. The clock attached to the muon has not ticked off a mean lifetime in the time these points travel past the muon.

These considerations lead to what is called the "twin paradox." Identical twins O and O' are separated at birth. O remains fixed, while O' moves away at high speed, and then returns to the birth position. Each sees her sister travel away at high speed, and then return to the position of birth. Due to time dilation, you might think that each twin is correct in saying, "I am younger." For when time comparison was considered above, if O' considered clock B', the time on B' was less than on A, while if O considered clock A, the time on A was less than on B'.

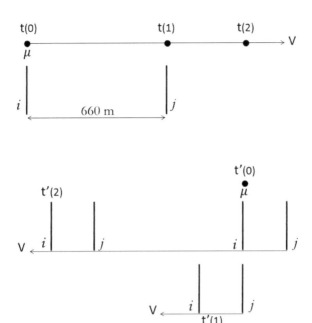

Fig. 1.4   Travel distance of speedy muons: top, bottom are the lab, muon views.

However, at least one twin experienced acceleration in order to turn around. An accelerating observer is not an inertial observer, and cannot always use the Lorentz transform of SR. The trips are asymmetric. Both twins seek to predict the trip time on the other's clock, and test the prediction, when they are again coincident. The simpler calculation, made by the inertial twin, is reviewed in Chapter 2. In Chapter 5, the accelerated twin's calculation will use the full power of GR.

1.4   Spacetime Diagrams

Many GR texts stress the concept of spacetime diagrams. Analytic calculations are favored by this author. However, for completeness, the former are briefly discussed in this section. One draws on a flat sheet, and must be concerned with four coordinates. Thus, one-dimensional motion, in the z-direction, is considered.

Draw a set of axes for frame O as shown on the top of Fig. 1.5. The upward vertical axis represents increasing time t, while the rightward

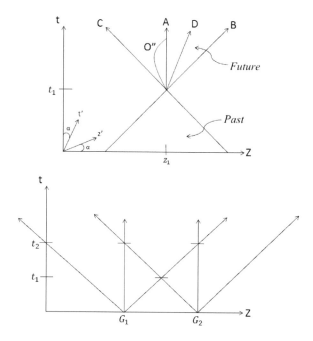

Fig. 1.5   Light cone physics, top for various travelers, bottom for separated galaxies.

horizontal axis represents increasing z. At some time t_1, particle A is at z_1. If particle A has rest mass, it can remain at rest with respect to O. Its world line is just a vertical line upward from (z_1, t_1). On the clock attached to A, the coordinate time duration dt is just the proper-time duration $d\tau$. Photons B and C, starting from the same position and time, must move with unit speed $|\frac{dz}{dt}| = 1$. Events outside of the triangle, defined by world lines B and C, are inaccessible starting from (z_1, t_1). This triangle is called the future triangle. If motion in two spatial directions is considered, the triangle becomes a cone.

A massive particle D, that moves with constant speed $V < 1$ relative to O, has a world line with $|\frac{dz}{dt}| < 1$. Its proper-time duration, measured relative to t_1, is $d\tau = dt/\gamma$. Here, dt is the time on a clock at rest with respect to A that, according to D, is coincident with D's position. Let twins O and O'' start at this spacetime point. O notes that her twin, moving with accelerated motion, always stays within the future triangle, at first moving away from O, but later returning. The gist of the twin "paradox" is that it is only O'' that experiences acceleration. If not created at (z_1, t_1), then

A, B, C, D, and O'' can get there only from the past triangle, obtained by extending B and C backwards in time.

Relative to the axes of O, the time and space axes (t', z') of frame O', in which D is at rest, can be drawn with hyperbolic angle $\alpha = \tanh^{-1} V$. This is because the Lorentz transform requires,

$$t = t' \cosh \alpha + z' \sinh \alpha = \gamma[t' + V z'],$$

$$z = z' \cosh \alpha + t' \sinh \alpha = \gamma[z' + V t'].$$

The light cones of two observers, in separated galaxies at $z_{1,2} = G_{1,2}$ are shown on the bottom of Fig. 1.5. Initially their light is non-interacting. Only after time t_1 will some light signals from the two observers be able to interact. Only after time t_2 will light signals originating from $G_{1,2}$ be able to form one side of a future triangle at $G_{2,1}$. Actually, the situation is much more complicated because the universe is expanding. Space is being created so that the distance between any pair of typical particles is increasing with time. Moreover, the expansion is accelerating. There may be an observer G_2, that observer G_1 sees separating faster than the speed of light. This is a phenomenon, discussed in a later chapter, that is not built into relativity.

Problems

1. Convert from MKS units to natural units where $G = c = 1$: (a) luminosity flux $= 10^{10} \, \mathrm{J\,s^{-1}\,m^{-2}}$; (b) density of water $= 10^3 \, \mathrm{kg\,m^{-3}}$. Convert from natural units to MKS units: rest energy density of a proton, $0.3 \times 10^{-9} \, \mathrm{m^{-2}}$.

2. Consider the rotations that transform the unit vectors, $\hat{e}_{x,y,z} \leftrightarrow \hat{e}_{\rho,\phi,z} \leftrightarrow \hat{e}_{\theta,\phi,r}$, into one another. Show that the spatial components of an arbitrary vector, written in terms of unit vectors, transform in the same way as the unit vectors.

3. Let the 3-vector in Problem 2 be $d\vec{r}$, with components $(dx, \, dy, \, dz)$ in either unit or basis vector notation. Use the transforms in Fig. 1.1 to find the components in cylindrical and spherical coordinates in terms of unit vectors. Use transforms similar to Eq. (1.4) to find the components in terms of basis vectors. What is the relation of the unit vectors to basis vectors in cylindrical and spherical coordinates?

4. O and O' have parallel axes. According to O, O' is moving with velocity,

$$\vec{V} = (V^1, V^2, V^3) = |\vec{V}|(\sin a \cos b, \sin a \sin b, \cos a).$$

What is the Lorentz transform $x^{\mu}{}_{,\nu'}$ between these reference frames? What is the relationship between $x^{\mu}{}_{,\nu'}$ and $x^{\nu'}{}_{,\mu}$? Prove that the proper time Eq. (1.8) is an invariant. If O looks at a rod that formally had rest length L that now is aligned with the velocity vector and moving with velocity $\vec{V}$, what will O consider the length to be?

Do this problem as follows: find the rotation that takes O' to O'' such that $\hat{e}^{3''}$ is aligned with $\vec{V}$. Then apply the Lorentz transform. This takes you to system $\bar{O}$, where $\bar{O}$ is obtained from O by the same rotation that took O' to O''. Now apply the inverse rotation to get from $\bar{O}$ to O.

5. Consider three inertial observers O, O', and O''. These observers could be considered the origins of three frames of reference with parallel rectangular axes. The clocks at rest relative to themselves are synchronized when they overlap. When the clock attached to O'' has ticked off a time $dx^{0''} = dt''$, calculate the time $dx^0 = dt$ and position information $dx^{i=2,3} = (dy, dz)$ of O'', according to O, for the following cases:

Case I: O'' moves with speed V^2 in the y-direction with respect to O', and O' moves with speed V^3 in the z-direction with respect to O.

Case II: O'' moves with speed V^3 in the z-direction with respect to O', and O' moves with speed V^2 in the y-direction with respect to O.

  (a) Calculate $\Delta f \equiv f_I - f_{II}$, where $f = dt$, dz, dy, $(dt)^2 - (dy)^2 - (dz)^2$ and $(dy)^2 + (dz)^2$.

  (b) Show that these results agree with those of Problem 4.

6. Cosmic rays are composed mainly of high-energy protons. Neglect gravity and assume the velocities of the protons are along a line passing through the center of earth. Let protons $(1, 2)$ move towards the earth's center with speeds $(V, \bar{V})$, relative to earth's center.

  (a) What is the speed of proton 2 relative to proton 1 if the velocities relative to earth's center are in the same or opposite direction? Determine these speeds if $V = 0.99$, $\bar{V} = 0.98$.

  (b) Suppose the protons are spaceships traveling in the same direction and at one time were at the same position, where they could synchronize their clocks. After a time T on the clock of the slower ship, it emits photons at intervals ΔT on its own clock. These photons travel to the faster ship. On the latter's clock, what is the interval between received photon bursts?

7. According to O, O' is moving with velocity $\vec{V} = 0.9\hat{e}_3$. In O', there are mirrors at $x^{3'} = \pm 1$ m. At $t = t' = 0$, light rays are emitted in O' heading for each of the mirrors. According to O, what are the positions and times that each mirror reflects the rays? What is the time and position of O', when the rays return to O'? Can O' place the mirror at $x^{3'} = -1$ m, somewhere else at negative $x^{3'}$, so that O says both mirrors reflect the light at the same time? According to O, will both rays return to $x^{3'} = 0$ at the same time?

8. Consider a runner holding a 10 m pole at its midpoint. The runner is at rest with respect to a barn, that is 5 m long. The runner gets to a speed V, to run through the barn, whose (rear, front) door is (closed, open). The speed is such that the pole appears to be 5 m long, according to a barn observer, at the barn's midpoint. The barn observer sees, at the instant the runner is at the midpoint of the barn, the pole just fits within the barn. At this instant, the rear door can be opened, the front door can be shut, and the pole passes through. The runner says the barn is only 2.5 m long. According to the runner, how does the pole avoid being hit by the doors? In this question, assume the doors open and close instantaneously.

9. High-energy muons are a component of cosmic rays. Suppose these muons are moving towards the earth's center with speed $v = 0.999$. According to earth clocks how long does it take for a trip of 1300 m? This is a trip down a moderately high mountain. How long does the trip take according to clocks attached to the muons? What fraction of the starting muons complete the trip? According to these muons, what is the distance traveled? Measurements at the top and bottom of such a mountain gave the first confirmation of time dilation.

Chapter 2

Vectors and Tensors in Spacetime

2.1 Metric Tensor

In Chapter 1, the contravariant displacement vector $dr^{\bar{\mu}}$ and the invariant proper-time element $d\tau$ were discussed. It was noted that $(d\tau)^2 \neq \sum_{\mu=0}^{3} dr^{\bar{\mu}} dr^{\bar{\mu}}$, but rather, from Eqs. (1.4) and (1.5), and the discussion following, summation occurs when the same index appears as both covariant and contravariant. Thus,

$$\begin{aligned} (d\tau)^2 &= (dr^{\bar{0}})^2 - [(dr^{\bar{1}})^2 + (dr^{\bar{2}})^2 + (dr^{\bar{3}})^2] \\ &= (dt)^2 - [(dx)^2 + (dy)^2 + (dz)^2] \\ &\equiv -dr_{\bar{\mu}} dr^{\bar{\mu}} \equiv -g_{\bar{\mu}\bar{\nu}} dr^{\bar{\nu}} dr^{\bar{\mu}}. \end{aligned} \qquad (2.1)$$

A vector necessarily has covariant and contravariant components. As shall be seen below, since $d\tau$ is an invariant and $dr^{\bar{\mu}}$, $dr_{\bar{\mu}}$ are vectors,

$$(d\tau)^2 = -dr_\mu dr^\mu = -g_{\mu\nu} dr^\nu dr^\mu \qquad (2.2)$$

$$= -g_{\nu\mu} dr^\mu dr^\nu = -g_{\nu\mu} dr^\nu dr^\mu. \qquad (2.3)$$

The quantity $g_{\mu\nu}$ with two covariant indexes is a tensor of rank 2 called the covariant metric tensor. By convention, when summing over an index that appears as both contravariant and covariant, the sum is over 0–3 for a Greek index and over 1–3 for a Roman index, e.g., $\delta^{\mu}{}_{\mu} = 4$ but $\delta^{i}{}_{i} = 3$. Equations (2.2) and (2.3) yield $g_{\mu\nu} = g_{\nu\mu}$, which is a symmetric tensor. In an inertial frame using rectangular coordinates, the metric tensor is particularly simple, and given a special symbol $g_{\bar{\mu}\bar{\nu}} = \eta_{\mu\nu}$. From Eq. (2.1), we have

$$1 = \eta_{ii} = -\eta_{00}, \quad \eta_{\mu\nu} = 0, \ \mu \neq \nu. \tag{2.4}$$

There is no need to put bars on the η indexes, they are understood.

From Eqs. (2.2) and (2.3), it is seen that the covariant metric tensor lowers indices, or converts a contravariant index into a covariant one. As the determinant of $g_{\mu\nu} \neq 0$, it must have an inverse — the contravariant metric tensor. The latter $g^{\mu\nu} = g^{\nu\mu}$ converts a covariant into a contravariant index,

$$dr^\mu dr_\mu = g_{\mu\nu} dr^\mu dr^\nu = g^{\mu\nu} dr_\mu dr_\nu,$$
$$dr^\nu = g^{\mu\nu} dr_\mu = g^{\mu\nu} g_{\mu\beta} dr^\beta,$$
$$g^{\mu\nu} g_{\mu\beta} = g^\nu{}_\beta = \delta^\nu{}_\beta. \tag{2.5}$$

The mixed form of the metric tensor is not part of current usage, however, it indicates that $\delta^\nu{}_\beta$ is a mixed tensor of rank 2.

In GR the metric tensor elements may be complicated functions of position and time. Once they are obtained, the problem is effectively solved. Staying in an inertial frame, it is easy enough to show that this tensor acquires complexity. One can just transform to cylindrical coordinates (see below).

2.2 Vector Transforms

So far we have seen that there are quantities that don't depend on an index and are invariants, for example, $d\tau$. In view of the discussion following Eq. (1.5), quantities that depend on one index, and transform as in Eq. (1.4), between different reference frames or coordinate systems, are contravariant vectors or tensors of rank 1. The transforms of tensors of higher rank are discussed in Section 2.3. Consider observers O and O′, with coordinates x^μ and $x^{\mu'}$, the components of contravariant vectors transform like,

$$V^\mu = x^\mu{}_{,\nu'} V^{\nu'} \quad \text{or} \quad g^{\mu\alpha} V_\alpha = x^\mu{}_{,\nu'} g^{\beta'\nu'} V_{\beta'}, \tag{2.6}$$

and

$$V_\sigma = \delta_\sigma{}^\alpha V_\alpha = g_{\sigma\mu} g^{\mu\alpha} V_\alpha$$
$$= g_{\sigma\mu} g^{\beta'\nu'} x^\mu{}_{,\nu'} V_{\beta'} = x^{\beta'}{}_{,\sigma} V_{\beta'}. \tag{2.7}$$

Equation (2.7) is the rule for transforming the components of covariant vectors. These results show that Eq. (2.1) leads to Eq. (2.2),

$$\begin{aligned}
(d\tau)^2 = dr_{\bar{\mu}}dr^{\bar{\mu}} &= (x^{\chi}{}_{,\bar{\mu}}\, dr_{\chi})(x^{\bar{\mu}}{}_{,\nu}\, dr^{\nu}) \\
&= x^{\chi}{}_{,\bar{\mu}}\, x^{\bar{\mu}}{}_{,\nu}\, dr_{\chi}dr^{\nu} = \delta^{\chi}{}_{\nu}dr_{\chi}dr^{\nu} \\
&= dr_{\nu}dr^{\nu}.
\end{aligned} \tag{2.8}$$

The metric tensor gaining complexity can be illustrated for O' being cylindrical and O being rectangular coordinates. The relations between coordinates are as follows:

$$\begin{aligned}
x &= \rho\cos\phi, \; y = \rho\sin\phi, \\
\rho &= (x^2 + y^2)^{1/2}, \; \phi = \tan^{-1}(y/x), \\
dx &= x_{,\rho}\, d\rho + x_{,\phi}\, d\phi = \cos\phi d\rho - \rho\sin\phi d\phi, \\
dy &= y_{,\rho}\, d\rho + y_{,\phi}\, d\phi = \sin\phi d\rho + \rho\cos\phi d\phi.
\end{aligned}$$

Thus,

$$(dx)^2 + (dy)^2 = (d\rho)^2 + \rho^2(d\phi)^2.$$

The position vector r^{μ} has components (t, x, y, z) and $r^{\mu'}$ has components $(t, \rho, 0, z)$. The displacement vector dr^{μ} has components (dt, dx, dy, dz) and $dr^{\mu'}$ has components $(dt, d\rho, d\phi, dz)$. Note, these components are in terms of basis vectors.

In an inertial frame,

$$\begin{aligned}
r_{\bar{\mu}} &= \eta_{\mu\nu}r^{\bar{\nu}}, \\
r_{\bar{0}} &= \eta_{00}r^{\bar{0}} = -r^{\bar{0}} = -t, \\
r_{\bar{i}} &= \eta_{ii}r^{\bar{i}} = r^{\bar{i}} = (x, y, z),
\end{aligned}$$

similarly,

$$dr_{\bar{\mu}} = \eta_{\mu\nu}dr^{\bar{\nu}} = (-dt, dx, dy, dz).$$

These results would be considerably more complicated in a non-inertial frame.

It is obvious that the t and z components are the same in both coordinate systems. Only the spatial components, in the plane perpendicular to z need to be transformed,

$$\begin{aligned}
r^{\bar{1}'} &= r^{\bar{1}'}{}_{,\bar{\mu}}\, r^{\bar{\mu}} = \rho_{,\bar{\mu}}\, r^{\bar{\mu}} = \rho_{,x}\, x + \rho_{,y}\, y = ((x)^2 + (y)^2)^{1/2} = \rho, \\
r^{\bar{2}'} &= r^{\bar{2}'}{}_{,\bar{\mu}}\, r^{\bar{\mu}} = \phi_{,\bar{\mu}}\, r^{\bar{\mu}} = \phi_{,x}\, x + \phi_{,y}\, y = \rho^{-2}(-yx + xy) = 0,
\end{aligned}$$

$$r_{\bar{1}'} = r^{\bar{\mu}},_{\bar{1}'}, r_{\bar{\mu}} = x^{\bar{\mu}},_{\bar{1}'}, r_{\bar{\mu}} = x,_\rho\, x + y,_\rho\, y = \rho^{-1}((x)^2 + (y)^2) = \rho = r^{\bar{1}'},$$
$$r_{\bar{2}'} = r^{\bar{\mu}},_{\bar{2}'}, r_{\bar{\mu}} = x^{\bar{\mu}},_{\bar{2}'}, r_{\bar{\mu}} = x,_\phi\, x + y,_\phi\, y = -yx + xy = 0.$$

This calculation is repeated for the displacement vector:

$$dr^{\bar{1}'} = \rho,_x\, dx + \rho,_y\, dy = \rho^{-1}(xdx + ydy) = d\rho,$$
$$dr^{\bar{2}'} = \phi,_x\, dx + \phi,_y\, dy = \rho^{-2}(-ydx + xdy) = d\phi,$$
$$dr_{\bar{1}'} = x,_\rho\, dx + y,_\rho\, dy = \rho^{-1}(xdx + ydy) = d\rho = dr^{\bar{1}'},$$
$$dr_{\bar{2}'} = x,_\phi\, dx + y,_\phi\, dy = (-ydx + xdy) = \rho^2 d\phi = \rho^2 dr^{\bar{2}'}.$$

Since $d\tau$ is an invariant, the metric is easily obtained:

$$\eta_{ij} dr^{\bar{i}} dr^{\bar{j}} = g_{\bar{i}'\bar{j}'} dr^{\bar{i}'} dr^{\bar{j}'},$$
$$(dx)^2 + (dy)^2 = (d\rho)^2 + \rho^2 (d\phi)^2$$
$$= g_{\rho\rho}(d\rho)^2 + g_{\phi\phi}(d\phi)^2 + 2g_{\rho\phi} d\rho d\phi,$$
$$g_{\bar{1}'\bar{1}'} = g_{\rho\rho} = 1,\ g_{\bar{2}'\bar{2}'} = g_{\phi\phi} = \rho^2,\ g_{\bar{1}'\bar{2}'} = 0,\ g^{\bar{i}'\bar{i}'} = 1/g_{\bar{i}'\bar{i}'}.$$

The reader should show that the following required relations hold:

$$(r_{\bar{\mu}'}, dr_{\bar{\mu}'}) = g_{\bar{\mu}'\bar{\nu}'}(r^{\bar{\nu}'}, dr^{\bar{\nu}'}),$$
$$(r^{\bar{\mu}'}, dr^{\bar{\mu}'}) = g^{\bar{\mu}'\bar{\nu}'}(r_{\bar{\nu}'}, dr_{\bar{\nu}'}),$$
$$g^{\bar{\mu}'\bar{\nu}'} dr_{\bar{\mu}'} dr_{\bar{\nu}'} = g_{\bar{\mu}'\bar{\nu}'} dr^{\bar{\mu}'} dr^{\bar{\nu}'}.$$

In Problem 4, the reader is led through the proof that the invariant 4-volume is given by

$$d^4 V = (-\det(g_{\mu\nu}))^{1/2} dx^0 dx^1 dx^2 dx^3. \tag{2.9}$$

The minus sign insures that the argument of the square root is positive. If just spatial parts are considered, it is not needed. The value of the spatial part in an inertial frame is $dxdydz$ for rectangular coordinates and $\rho d\rho d\phi dz$ for cylindrical coordinates. The case of spherical coordinates is left as an exercise. Then in any GR frame, for a one-dimensional distance, say between radial coordinates (a, b), the proper distance is as follows:

$$L_p = \int_a^b dr(g_{rr})^{1/2}. \tag{2.10}$$

It depends on the metric. Generally, the inertial frame value is $|b-a|$. Where gravity acts, it typically is larger because spacetime is curved. The speed of light is still unity because the proper time, that on a clock at rest with respect to an observer, runs slower in stronger gravity.

The spatial parts of vectors are usually written in terms of components V^i and orthogonal basis vectors $\vec{e}_i$. In rectangular coordinates, the latter can be taken as constant unit vectors $\hat{e}_i, i = x, y, z$. In cylindrical coordinates, the results of Problem 1.3 gave $\vec{e}_\rho = \hat{e}_\rho$ and $\vec{e}_\phi = \rho \hat{e}_\phi$. They are not constant, but vary with ϕ. The correct distance squared between points in the ρ, ϕ plane, in an inertial system, is obtained from $|d\rho \hat{e}_\rho + d\phi \rho \hat{e}_\phi|^2$. However, it is not necessary to use basis vectors when we work with the metric tensor.

In Chapter 1, the Lorentz transform between frames O and O$'$, in relative motion, was studied. The transform for $dr^{\bar{\mu}}$ and $dr_{\bar{\mu}}$ was just Eqs. (2.6) and (2.7). So quantities depending on one index, and transforming via the Lorentz transform, are vectors in spacetime. This property will be used below to construct additional vectors.

2.3 Tensor Transforms

To study the transformation properties of tensors of rank > 1, start by constructing a quantity that depends on two indexes that have a known transformation, e.g., $V^\mu W_\nu$, $V^\mu W^\nu$ and $V_\mu W_\nu$, where V and W are vectors. Such quantities are defined to transform as tensors, and it is easy to see how higher ranked tensors must transform. Using Eqs. (2.6) and (2.7), we have

$$(V^\mu W_\nu) = x^\mu{}_{,\psi'} \, V^{\psi'} x^{\xi'}{}_{,\nu} \, W_{\xi'} = x^\mu{}_{,\psi'} \, x^{\xi'}{}_{,\nu} \, (V^{\psi'} W_{\xi'}), \qquad (2.11)$$

$$(V^\mu W^\nu) = x^\mu{}_{,\psi'} \, V^{\psi'} x^\nu{}_{,\xi'} \, W^{\xi'} = x^\mu{}_{,\psi'} \, x^\nu{}_{,\xi'} \, (V^{\psi'} W^{\xi'}), \qquad (2.12)$$

$$(V_\mu W_\nu) = x^{\psi'}{}_{,\mu} \, V_{\psi'} x^{\xi'}{}_{,\nu} \, W_{\xi'} = x^{\psi'}{}_{,\mu} \, x^{\xi'}{}_{,\nu} \, (V_{\psi'} W_{\xi'}). \qquad (2.13)$$

One then declares that the above multiplications of the two vectors produce tensors of rank 2, such as $T^\mu{}_\nu$, $T^{\mu\nu}$, and $T_{\mu\nu}$. The above equations are the transformation rule, for any tensor of rank 2. However, this must be tested on two quantities asserted to be tensors of rank 2, the metric and Kronecker delta tensors. Let V, W be vectors. If $g_{\mu\nu}$ is a tensor, then

$$\begin{aligned}
g_{\mu\nu} V^\mu W^\nu &= x^{\xi'}{}_{,\mu} \, x^{\chi'}{}_{,\nu} \, g_{\xi'\chi'} x^\mu{}_{,\alpha'} \, V^{\alpha'} x^\nu{}_{,\beta'} \, W^{\beta'} \\
&= x^{\xi'}{}_{,\mu} \, x^\mu{}_{,\alpha'} \, x^{\chi'}{}_{,\nu} \, x^\nu{}_{,\beta'} \, g_{\xi'\chi'} V^{\alpha'} W^{\beta'} \\
&= x^{\xi'}{}_{,\alpha'} \, x^{\chi'}{}_{,\beta'} \, g_{\xi'\chi'} V^{\alpha'} W^{\beta'} \\
&= \delta^{\xi'}{}_{\alpha'} \delta^{\chi'}{}_{\beta'} g_{\xi'\chi'} V^{\alpha'} W^{\beta'} = g_{\xi'\chi'} V^{\xi'} W^{\chi'}.
\end{aligned}$$

This result is an invariant. It agrees with Eq. (2.3) when $V^\mu = W^\mu = dr^\mu$. If $g_{\mu\nu}$ was not a tensor, this result wouldn't hold.

The transform of the Kronecker tensor is left as a problem. Here are some explicit examples, of expected results, between rectangular and cylindrical coordinates:

$$g_{\rho\rho} = (x_{,\rho})^2 \eta_{xx} + (y_{,\rho})^2 \eta_{yy} = \cos^2\phi + \sin^2\phi = 1,$$
$$g_{\phi\phi} = (x_{,\phi})^2 \eta_{xx} + (y_{,\phi})^2 \eta_{yy} = \rho^2(\sin^2\phi + \cos^2\phi) = \rho^2,$$
$$g^{\rho\rho} = (\rho_{,x})^2 \eta^{xx} + (\rho_{,y})^2 \eta^{yy} = (x^2 + y^2)/\rho^2 = 1,$$
$$g^{\phi\phi} = (\phi_{,x})^2 \eta^{xx} + (\phi_{,y})^2 \eta^{yy} = (y^2 + x^2)/\rho^4 = 1/\rho^2,$$
$$\delta^\rho_{\ \rho} = \rho_{,x}\, x_{,\rho}\, \delta^x_{\ x} + \rho_{,y}\, y_{,\rho}\, \delta^y_{\ y} = \cos^2\phi + \sin^2\phi = 1,$$
$$\delta^\phi_{\ \phi} = \phi_{,x}\, x_{,\phi}\, \delta^x_{\ x} + \phi_{,y}\, y_{,\phi}\, \delta^y_{\ y} = \sin^2\phi + \cos^2\phi = 1,$$
$$\delta^\rho_{\ \phi} = \rho_{,x}\, x_{,\phi}\, \delta^x_{\ x} + \rho_{,y}\, y_{,\phi}\, \delta^y_{\ y} = (-xy + yx)/\rho = 0.$$

One can also check that under a Lorentz transformation the metric tensor transforms correctly. One obtains, using Eqs. (1.9)–(1.11),

$$\eta^{00} = x^{\bar{0}}_{\ ,\bar{\alpha}'}\, x^{\bar{0}}_{\ ,\bar{\beta}'}\, \eta^{\alpha'\beta'}$$
$$= (x^{\bar{0}}_{\ ,\bar{0}'})^2 \eta^{0'0'} + (x^{\bar{0}}_{\ ,\bar{3}'})^2 \eta^{3'3'} = \gamma^2(-1 + V^2) = -1.$$

The other elements can similarly be shown to work out. Try it.

It is now obvious that a tensor of rank integer n transforms such that for each contravariant index there will be a partial derivative factor as in Eq. (2.6) and for each covariant index, a partial derivative factor as in Eq. (2.7). The product of the factors multiplies an element of the tensor, e.g.,

$$T^{\alpha\beta}_{\ \ \gamma\delta} = x^\alpha_{\ ,\mu'}\, x^\beta_{\ ,\nu'}\, x^{\xi'}_{\ ,\gamma}\, x^{\psi'}_{\ ,\delta}\, T^{\mu'\nu'}_{\ \ \xi'\psi'}. \qquad (2.14)$$

2.4 Forming Other Vectors

Since dr^μ is a vector and $d\tau$ is an invariant, the quantity $U^\mu = \frac{dr^\mu}{d\tau}$ is another vector with units of velocity. This is not true for photons because $d\tau = 0$. Suppose SR observer O′ says an object has 3-velocity rectangular components $\frac{dr^{i'}}{dt'} = v^{i'} = v_{i'}$. An observer moving with the object says that $d\tau$ has elapsed. Chapter 1 results yielded $d\tau = dt'/\bar{\gamma}'$, where dt' is the time elapsed according to O′. Here $\gamma' = (1 - |v'|^2)^{-1/2}$ and $|v'|^2 = v^{\bar{i}'} v_{\bar{i}'}$. Then,

$$U^{\bar{i}'} = \gamma' \frac{dr^{\bar{i}'}}{dt'} = \gamma' v^{\bar{i}'} = U_{\bar{i}'}, \tag{2.15}$$

$$U^{\bar{0}'} = \gamma' \frac{dr^{\bar{0}'}}{dt'} = \gamma' = -U_{\bar{0}'}, \tag{2.16}$$

$$-\eta_{\mu'\nu'} U^{\bar{\mu}'} U^{\bar{\nu}'} = -\gamma'^2(-1 + |v'|^2) = 1, \quad \text{invariant.} \tag{2.17}$$

There are similar equations for the $U^{\bar{\mu}}$ in O — just remove the primes.

The 3-velocity transforms come from applying the Lorentz transform to these vectors. If O′ moves relative to O, with speed V in the z direction, let $\gamma[V] = (1 - |V|^2)^{-1/2}$. Then,

$$U^{\bar{0}} = \gamma = \gamma[V](U^{\bar{0}'} + VU^{\bar{3}'}) = \gamma[V]\gamma'(1 + Vv^{\bar{3}'}), \tag{2.18}$$

$$U^{\bar{3}} = \gamma v^{\bar{3}} = \gamma[V](U^{\bar{3}'} + VU^{\bar{0}'}) = \gamma[V]\gamma'(v^{\bar{3}'} + V),$$

$$v^{\bar{3}} = \frac{v^{\bar{3}'} + V}{1 + Vv^{\bar{3}'}} = \frac{\gamma[V](dx^{\bar{3}'} + V dx^{\bar{0}'})}{\gamma[V](dx^{\bar{0}'} + V dx^{\bar{3}'})} = \frac{dx^{\bar{3}}}{dt}, \tag{2.19}$$

$$\gamma v^{\bar{1},\bar{2}} = \gamma' v^{\bar{1}',\bar{2}'},$$

$$v^{\bar{1},\bar{2}} = \frac{v^{\bar{1}',\bar{2}'}}{\gamma[V](1 + V\bar{v}^{\bar{3}'})} = \frac{dx^{\bar{1}',\bar{2}'}}{\gamma[V](dx^{\bar{0}'} + V dx^{\bar{3}'})} = \frac{dx^{\bar{1},\bar{2}}}{dt}. \tag{2.20}$$

The quantity $d\tau$ has vanished from Eqs. (2.19) and (2.20). Thus, they also hold for photons, e.g., if $v^{\bar{1}'} = 1$, $v^{\bar{2}'} = v^{\bar{3}'} = 0$, then $v_{\bar{i}} v^{\bar{i}} = (\gamma[V])^{-2}$, $v^{\bar{2}} = 0$, $v_{\bar{3}} v^{\bar{3}} = |V|^2$ and $|v|^2 = v_{\bar{i}} v^{\bar{i}} = 1$. All observers see the same speed for light. For a slowly moving object like the earth, under the gravitational influence of the sun, $\gamma \approx 1$, $U^0 \approx 1$ and $U^i \approx 0$.

2.5 Twin Problem Revisited

The twin problem with a simple acceleration can now be considered. Twin O is at rest, while twin O′ is positioned in a rocket, as shown in Fig. 2.1. The rocket accelerates in the z-direction such that O′ experiences an acceleration $g = 9.807/(9 \times 10^{16})\,\mathrm{m}^{-1}$ and is quite comfortable. On the clock of O′ that measures the proper time of O′, the acceleration lasts for a time T. The acceleration reverses direction for a time $2T$, and then reverses direction again for a time T. This brings the rocket back to rest, and O′ to the position of O. What is the time on the clock of O? O can interpret the problem with SR.

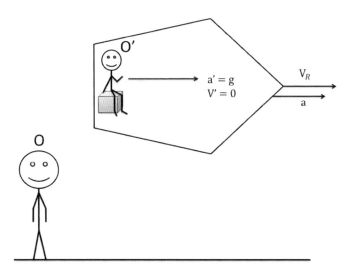

Fig. 2.1 The twin problem: O' at rest in the rocket experiences acceleration of magnitude g on going from, and returning to twin O.

The 3-acceleration magnitude of O' is g according to O'. The 3-acceleration of O' according to O must be calculated. The latter is $\frac{dV_R}{dt}$, where V_R is the speed of the rocket with respect to O. The 3-acceleration enables the calculation of the rocket's speed with respect to O. That allows comparison of clocks. The 3-acceleration is obtained from an acceleration vector $A^\mu \equiv \frac{dU^\mu}{d\tau}$ similar to the way the 3-velocity was obtained from U^μ. In general, for motion in O, in the 3-direction with speed V and acceleration $a = \frac{dV}{dt}$,

$$A^{\bar\mu} \equiv \frac{dU^{\bar\mu}}{d\tau} = \gamma\frac{dU^{\bar\mu}}{dt}, \tag{2.21}$$

$$A^{\bar0} = \gamma\frac{d\gamma}{dt} = \gamma\frac{-1}{2(1-V^2)^{-3/2}}\left(-2V\frac{dV}{dt}\right)$$
$$= \gamma^4 aV, \tag{2.22}$$

$$A^{\bar3} = \gamma\frac{d}{dt}\frac{dx^{\bar3}}{d\tau} = \gamma\frac{d}{dt}\left(\gamma\frac{dx^{\bar3}}{dt}\right) = \gamma\frac{d(\gamma V)}{dt}$$
$$= \gamma\left(\gamma a + \frac{d\gamma}{dt}V\right) = \gamma^2 a + \gamma^4 V^2 a$$
$$= \gamma^2 a(1+V^2\gamma^2) = \gamma^4 a. \tag{2.23}$$

The rocket is not an inertial frame, but at each instant looks like one. It is moving with speed V_R relative to O. O' is at rest relative to the rocket, but experiences $A^{3'} = \pm g$. O can use the Lorentz transform to transform vectors from the rocket frame to the O frame,

$$A^{\bar{0}} = \gamma_R[A^{0'} + V_R A^{3'}], \qquad \gamma_R = (1 - V_R^2)^{-1/2},$$

$$V_R a \gamma_R^4 = \gamma_R V_R a' = \gamma_R V_R (\pm g),$$

$$a = \frac{dV_R}{dt} = \pm g \gamma_R^{-3}, \text{ so,} \tag{2.24}$$

$$\pm g \, dt = dV_R \gamma_R^3 = dV_R (1 - V_R^2)^{-3/2}.$$

Using the relationship between coordinate time and proper time yields,

$$(d\tau)^2 = (dt)^2 - (dz)^2 = (dt)^2 (1 - V_R^2),$$

$$d\tau = dt(1 - V_R^2)^{1/2} = (\pm 1/g) dV_R/(1 - V_R^2).$$

On the first leg $\tau = T - 0 = T$, take the plus sign, and integrate from $V_R = 0$ to $V_R(F)$. On the second leg $\tau = 2T - T = T$, take the minus sign, but integrate from $V_R = V_R(F)$ to 0. Thus the two legs yield the same result. For the third leg $\tau = 3T - 2T = T$, take the minus sign, but integrate from $V_R = 0$ to $-V_R(F)$. For the last leg $\tau = 4T - 3T = T$, take the plus sign, but integrate from $V_R = -V_R(F)$ to 0. Thus all legs yield the same result. So the calculation is made for the first leg:

$$T = (1/g) \int_0^{V_R(F)} dV_R/(1 - V_R^2) = (1/2g) \ln \frac{1 + V_R(F)}{1 - V_R(F)},$$

$$\exp(2gT) = \frac{1 + V_R(F)}{1 - V_R(F)}, \tag{2.25}$$

$$V_R(F) = \frac{\exp(2gT) - 1}{\exp(2gT) + 1} = \tanh gT \leq 1.$$

Then, from Eq. (2.24),

$$gt = \int_0^{V_R(F)} dV_R (1 - V_R^2)^{-3/2} = V_R(F)(1 - V_R(F)^2)^{-1/2},$$

$$= \sinh gT. \tag{2.26}$$

If $T = 1$ y $= 0.316 \times 10^8$ s $= 0.948 \times 10^{16}$ m, $gT = 1.033$, then $V_R(F) = \tanh 1.033 = 0.775$, $gt = \sinh 1.033 = 1.226$. This yields $t = (0.918 \times 10^{16})(1.226)$ m $= 0.375 \times 10^8$ s $= 1.187$ y. So, for the entire trip, O' ages 4 y, while O ages 4.748 y. Even more striking, if $T = 10$ y, then $t = 1.483 \times 10^4$ y!

2.6   Momentum and Energy

Another important vector comes from multiplying the velocity vector by an invariant quantity with units of mass $mU^\mu \equiv P^\mu$, where P is the momentum. In the system where $c = 1$, it has units of kg. In the natural units, where $c = G = 1$, it has units of m. By going to the rest frame of the object, one notes m is the rest mass. For an inertial observer, the spatial part of the momentum vector is the linear momentum. Its components are as follows:

$$P^{\bar{i}} = (1 - |\vec{V}|^2)^{-1/2} m V^{\bar{i}} = \gamma m V^{\bar{i}}, \quad |\vec{V}|^2 = V^{\bar{i}} V_{\bar{i}}. \qquad (2.27)$$

These reduce to their usual non-relativistic values when $|\vec{V}| \ll 1$. The time component is as follows:

$$P^{\bar{0}} = \gamma m \equiv E \qquad (2.28)$$
$$= m(1 + |\vec{V}|^2/2 + \cdots).$$

At low speeds, the last equation is the sum of the rest energy, and the non-relativistic kinetic energy. So, for any speed, the time component of the momentum vector is the sum of the rest energy and the relativistic kinetic energy KE. This is the total energy E. A non-invariant relativistic mass $M \equiv E/c^2 = E$ is a useful construct for the example in Section 2.8. A little manipulation yields other important relations:

$$m^2 = -g_{\mu\nu} P^\mu P^\nu, \text{ then for inertial observers}$$
$$= (\gamma m)^2 - P_{\bar{i}} P^{\bar{i}} = E^2 - |\vec{P}|^2, \qquad (2.29)$$
$$E = [|\vec{P}|^2 + m^2]^{1/2} \equiv KE + m.$$

Photons have $m = 0$, so for them,

$$E = KE = |\vec{P}| = M = h\nu, \qquad (2.30)$$

where ν is the frequency.

2.7   Doppler Shift

The Doppler shift for photons in SR can be obtained from the Lorentz transform of photon momentum. The situation is illustrated in Fig. 2.2. Frame O′ is moving relative to frame O with speed V in the z-direction. A laser, at rest in O′, emits photons with $P^{\bar{0}'} = E' = h\nu'$ and $P^{\bar{3}'} = -h\nu'$.

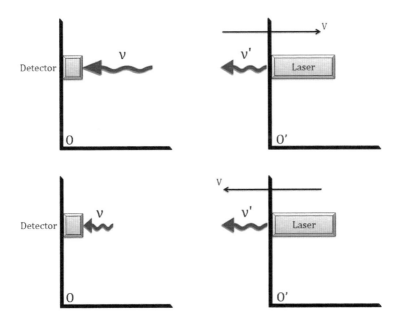

Fig. 2.2   The laser at rest in O' emits light of frequency ν'. The light received by a detector at rest in O measures frequency ν.

If the photons don't travel in the $-z$-direction, they won't get to the origin of O. From the Lorentz transform,

$$E = h\nu = \gamma h\nu'(1 - V),$$

$$\frac{\nu}{\nu'} = \frac{1 - V}{(1 - V^2)^{1/2}} = \left(\frac{1 - V}{1 + V}\right)^{1/2} = \frac{\lambda'}{\lambda}. \tag{2.31}$$

For a source receding from O, the received frequency is shorter, the wavelength is longer — a red shift. If you change the sign of V, so that the source is advancing, there is a blue shift. Examination of the line spectra from very faraway galaxies, and knowledge of the Doppler shift, leads to the conclusion that they are all separating from us — an expanding universe. One might think that the expansion is taking place in an already existing space, but we now know the expansion is due to more space being created between any pair of objects.

Another way to get this result is, perhaps more satisfying, as it uses invariants. In O', the light source is at rest, and the light has $P^{\bar{0}'} = h\nu'$, $P^{\bar{3}'} = -h\nu'$. O is moving in the $-z$-direction with speed V, with respect to O′. The velocity of O according to O′ is as follows:

$$U^{\bar{0}'}(O) = \gamma(U^{\bar{0}}(O) - VU^{\bar{3}}(O)) = \gamma U^{\bar{0}}(O) = \gamma,$$

$$U^{\bar{3}'}(O) = \gamma(U^{\bar{3}}(O) - VU^{\bar{0}}(O)) = -\gamma V U^{\bar{0}}(O) = -\gamma V,$$

$$\eta_{\mu'\nu'}U^{\bar{\mu}'}(O)P^{\bar{\nu}'}(\text{light}) = \gamma(-h\nu')(1 - V) = -h\nu'\left(\frac{1-V}{1+V}\right)^{1/2} = -E,$$

$$E = P^{\bar{0}}(O) = -U_{\bar{\mu}'}(O)P^{\bar{\mu}'} = -U^{\bar{\mu}'}(obs)P_{\bar{\mu}'}.$$

$$(2.32)$$

Here, E is the energy of the photons measured by the detector fixed in O; $U_{\bar{\mu}'}(O)$ is the velocity of O determined by O′; and $P^{\bar{\mu}'}$ is the momentum of the photons determined by O′. In the above case, if $V = 0$, then $U_{\bar{\mu}'}(O) = (-1, 0, 0, 0)$ and $E = -(-1)P^{\bar{0}'} = h\nu'$. For massive particles, you can easily show that Eq. (2.32) still holds.

2.8 Gravity Affects Time

Using the relativistic mass, it is easy to show that gravity affects time. Consider a photon with $KE = M = h\nu$, emitted at the surface of the sun, with mass M_s and radius R_s. When it gets far from the sun, say at earth, with mass M_e, radius R_e, and distance from the sun D, the frequency changes so that total energy is conserved. At earth, $KE' = M' = h\nu'$. The setup is illustrated in Fig. 2.3. The frequency of the sunlight at earth

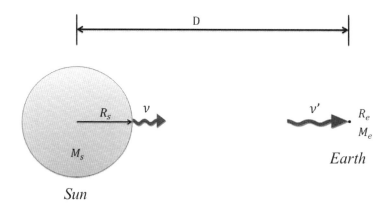

Fig. 2.3 Light of frequency ν is emitted at the surface of the sun. The light received at earth, where gravity is weaker, has frequency ν'.

can be calculated from energy conservation with Newtonian gravitational potential energy included. This calculation is not relativistically rigorous as it mixes non-relativistic and relativistic concepts. However, it gives the correct result when gravity is a very weak effect,

$$KE + PE = KE' + PE',$$
$$M - MM_s/R_s = M' - M'M_s/D - M'M_e/R_e,$$
$$h\nu(1 - M_s/R_s) = h\nu'(1 - [M_s/R_s][(R_s/D) + (M_e/M_s)(R_s/R_e)]),$$
$$R_s/D = 4.64 \times 10^{-3}, \quad (M_e/M_s)(R_s/R_e) = 0.33 \times 10^{-3},$$
$$1 - \nu'/\nu \approx M_s/R_s = 1.476 \times 10^3/6.96 \times 10^8 = 2.1 \times 10^{-6}.$$

So, at the sun, light is emitted by a physical process, like the line spectrum of helium. At earth, observers compare this light, with light from the line spectrum of helium, emitted on earth. The light from the sun has a lower frequency and larger wavelength than the earth light. This is the gravitational red shift. Frequency is inversely related to time; in this case, the time between successive wave crests. So the proper-time earth clock indicates the time between crests, of the light from the sun is longer, than the similar time from the earth light. This means it is longer than the time on a proper-time clock at the sun. The reason is, if a source is to be considered helium, it has to be helium everywhere. Otherwise, the first postulate of relativity is violated. The time between crests measured on a clock at the same position as the source doesn't depend on the position, regardless of gravity. Source process times and clock times are affected equally. Proper-time clocks in stronger gravity are ticking at a slower rate than those in weaker gravity.

An observer far from a black hole would say it takes an infinitely long time for a foolish astronaut to get to the horizon, whereas the astronaut doesn't think anything is amiss time-wise.

2.9   The Pound–Rebka Experiment

These predictions have been confirmed for light from stars, but was first confirmed in the earth's gravity, (see Pound and Rebka, 1959). The experiment required use of the Mössbauer effect and a wide variety of other physical processes.

As an introduction to this important experiment, the emission and absorption of photons are reviewed. Consider a free atom at rest, such as one in a gas with a nucleus in an excited energy state E_H. It will emit a photon and undergo a transition to a lower energy state E_L. In order to conserve energy and momentum, the atom must recoil,

$$E_H - E_L \equiv \Delta E = h\nu + |\vec{P}|^2/(2M_A) = h\nu + (h\nu)^2/(2M_A),$$

where ν is the photon frequency and M_A is the atomic mass. If a similar atom's nucleus was excited by absorbing a photon, then

$$\Delta E = h\nu' - |\vec{P'}|^2/(2M_A) = h\nu' - (h\nu')^2/(2M_A).$$

So $h\nu < \Delta E < h\nu'$. The emitted photon cannot be reabsorbed by another free nucleus, if as is the case here, the intrinsic line-width of the transition is small enough. The intrinsic line-width is due to the time–energy uncertainty relation. If the lifetime of the excited state is long, so that there is ample time to make an energy measurement, then the error in the energy will be very small.

In 1958, R. Mössbauer discovered that when such atoms were part of a crystal lattice, the crystal as a whole, with zero excitation of phonons, recoiled in a large fraction of the transitions. For this discovery, Mössbauer was awarded the Nobel prize. In this case, $M_A \to \infty$ and $h\nu = h\nu'$. Thus, the emitted photon could be reabsorbed. A very good source with which to see this effect is ^{57}Fe. It results in an excited nuclear state 14.4 keV above the ground state, from electron capture in ^{57}Co (lifetime 272 d). You can get the cobalt at a cyclotron, using the reaction ^{56}Fe(d, n)^{57}Co. The cobalt diffuses into the lattice when the iron target slab is heated. As natural iron has only 2.17% of ^{57}Fe, the slab can be enriched in isotope 57.

In a simple experiment, a thin enriched source slab is placed close to a thin absorber slab. Directly behind the latter is a photon detector, perhaps a NaI crystal mounted on a phototube. If the absorbing slab is ordinary iron, almost no reduction in counting rate is observed, but if it is enriched in ^{57}Fe, an obvious reduction is seen. In the latter case, more emitted photons are absorbed. This holds whether the experiment is conducted at the bottom or top of a high tower $H = 22.55$ m, in the case of the Pound–Rebka experiment. That is, if source and detector are at the same position, the measured frequency is independent of where they are.

In the Pound–Rebka experiment, the source was carefully prepared with a very narrow intrinsic line width. Let it be at the bottom $r_2 = R_e$, while a similarly prepared absorber is at the top $r_1 = R_e + H$, where gravity is weaker. After solid angle effects are accounted for, a reduction in count rate is not observed nor is one expected. The gravitational red shift reduces the photons' frequency so they cannot be reabsorbed. This confirms that the frequency at the top is not that at the bottom, but hasn't confirmed whether or not the frequency has decreased, nor by how much.

In order to measure this frequency change, and see if it agrees with the GR prediction, the source was set in motion. Then in addition to the above effect, there is also a Doppler shift. If the source moves towards the absorber with speed V, the frequency at the absorber is, using both the Doppler shift and the gravitational red shift,

$$\frac{\nu_{2,1}}{\nu_{2,2}} = \left(\frac{1+V}{1-V}\right)^{1/2} \left[1 - M_e \left(\frac{1}{r_2} - \frac{1}{r_1}\right)\right]$$
$$\approx (1+V) \left[1 - M_e \left(\frac{1}{R_e} - \frac{1}{R_e + H}\right)\right].$$

The approximation is due to $V \ll 1$. Since $H \ll R_e$ the above can be approximated as

$$\frac{\nu_{2,1}}{\nu_{2,2}} \approx (1+V) \left[1 - \frac{M_e}{R_e}\left(1 - \frac{1}{1 + H/R_e}\right)\right]$$
$$\approx (1+V)\left[1 - \frac{M_e}{R_e}(1 - (1 - H/R_e))\right]$$
$$= (1+V)\left[1 - \frac{M_e H}{R_e^2}\right].$$

For maximum absorption, one desires the above ratio to be unity, so that $1 + V = \left(1 - M_e H/R_e^2\right)^{-1} \approx 1 + M_e H/R_e^2$, or $V = 2.46 \times 10^{-15}$. In MKS units, this is 0.74×10^{-6} m/s. Moving the source so slowly would be extremely difficult. However, Pound and Rebka moved it in a sinusoidal manner, so that $V = V_0 \cos \omega t$. The speed V_0 could be much larger than the above value, but by observing the way the detector intensity varied with $\cos \omega t$, the red shift value could be deduced. The original experiment found agreement with GR to the 10% level. It has since been improved to less than 1%.

2.10   Global Positioning System

The global positioning system is in wide use, and GR provides an important correction. Before worrying about the role played by GR, let's look at a simple-minded system where only SR needs to be worried about.

Imagine our observer does not know where he is, though he is actually at dz_0. Overhead at height $h \ll R_e$, many planes fly with speed V in the z-direction. The planes are at rest with respect to each other. At high frequency ν, they broadcast their time and position (dt', dz'). These are connected to times and positions in the observer's frame by the Lorentz transform,

$$\gamma = (1 - V^2)^{-1/2} \approx 1 + V^2/2,$$
$$dt = \gamma(dt' + V\,dz'), \qquad dz = \gamma(dz' + V\,dt').$$

The time of flight dT of the electromagnetic wave to the observer is $dT = (h^2 + [dz - dz_0]^2)^{1/2}$, and the time of arrival at the observer is $dt_0 = dT + dt$. The observer has a system that interprets the signals so that two simultaneously arriving signals are deduced. Of course, this means that the two signals were not emitted simultaneously by the planes. Also simultaneous arrival really means arrival within a narrow time window, and that depends on the accuracy of your device. If two planes provide simultaneous arrivals, then

$$dT_A + dt_A = dT_B + dt_B,$$
$$(h^2 + [dz_A - dz_0]^2)^{1/2} + dt_A = (h^2 + [dz_B - dz_0]^2)^{1/2} + dt_B.$$

The above equation allows calculation of dz_0. The accuracy of the position depends on the accuracy of the time measurements. For extreme accuracy, the correction due to the relativistic factor γ needs to be included.

The real GPS, illustrated in Figs. 2.4 and 2.5, is a system of 24 satellites. They orbit earth with a period $T = 12\,\mathrm{h}$. Newton's theory will give us an idea of the accuracy and the corrections needed. Circular motion is considered:

$$mV^2/R = mM_e/R^2,$$
$$V = (M_e/R)^{1/2} = 2\pi R/T, \qquad R = [TM_e^{1/2}/(2\pi)]^{2/3},$$
$$T = (12)(3600)(3 \times 10^8)\,\mathrm{m},$$
$$M_e = (5.97 \times 10^{24})(0.742 \times 10^{-27})\,\mathrm{m},$$
$$R = 0.265 \times 10^5\,\mathrm{km} = 4.15 R_e,$$
$$V = 0.128 \times 10^{-4}.$$

The time correction has two parts: a factor γ from SR, and a factor from GR because the clocks of the receiver and satellite experience gravity of

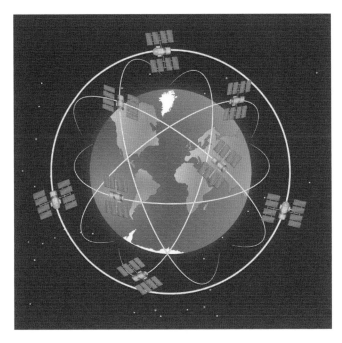

Fig. 2.4   Twenty-four GPS satellites orbit earth, and broadcast their positions and times at high frequency.

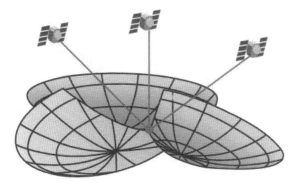

Fig. 2.5   Light speed signals from three satellites, that arrive at the same time, provide a unique earth position.

differing strength. The sizes of the corrections are as follows:

$$\gamma = (1 - V^2)^{-1/2} \approx 1 + V^2/2 = 1 + 1.6 \times 10^{-10},$$
$$(1 - M_e/R_e)/dt_E = (1 - M_e/(4.15R_e))/dt_S,$$
$$dt_e/dt_S = [1 - M_e/R_e]/[1 - M_e/(4.15R_e)]$$
$$\approx [1 - M_e/R_e][(1 + M_e/(4.15R_e)]$$
$$= 1 - 0.76(M_e/R_e) = 1 - 4 \times 10^{-10}.$$

The magnitude of the GR correction is 2.5 times larger than that from SR.

Motion about the sun affects earth and satellite equally, but you might think that the formula for γ could only be used at the poles, where an earth observer is at rest with respect to the earth's center. At other latitudes, an earth observer is moving in a circle about an axis through the earth's poles and center. However, the earth is not a sphere, and the polar radius is less than the equatorial radius. There is a latitude effect for the gravitational potential at the surface. You also must add to the brew that the earth is spinning, but one cannot tell the difference between the centripetal force and an additional oppositely directed gravitational force. The total gravitational potential at earth's surface turns out to be independent of latitude. So clocks anywhere on the surface run at the same rates. For a more complete discussion see Drake (2006).

Due to the large distance between the surface of earth and the satellite, exquisite timing accuracy is required to get a location accuracy of 2 m at earth's surface. This was the accuracy required by the military, e.g., in 1 ns light travels 0.3 m. So approximately 6 ns accuracy is needed. Modern atomic clocks can easily keep time to this accuracy. However, if both SR and GR corrections were not included, the errors would quickly compound, and the system wouldn't work. This alone makes GR very practical. It may be that a very clever person would have noticed that without the GR correction the GPS could be made to work by an arbitrary scaling of time. So we have heard it argued, that one does not have to know about GR to make the GPS work. However, without knowledge of GR, it would have taken a much longer time to get the GPS running correctly. Also, it is always better to understand a complex system rather than making use of arbitrary fixes.

There are other corrections to be mindful of: the relativistic Doppler effect, the varying of the index of refraction along the path of the electromagnetic wave, and that the orbits may not be exactly circular.

2.11   Tensor Equations

In a fully relativistic theory, energy is just one component of the momentum vector. Momentum and energy conservation is just conservation of each component of that vector. Suppose it holds in one frame where, for example, a muon, as illustrated in Fig. 2.6, decays into three lighter particles $\mu \rightarrow \nu\bar{\nu}e$. In the muon rest frame O, conservation of momentum is as follows:

$$0 = P^{\bar{\beta}}(I) - P^{\bar{\beta}}(F) = P^{\bar{\beta}}(\mu) - P^{\bar{\beta}}(\nu) - P^{\bar{\beta}}(\bar{\nu}) - P^{\bar{\beta}}(e). \quad (2.33)$$

The Lorentz transform is applied to every term in the above equation. This yields the momenta in a frame O', where the muon is moving. It is obvious that momentum conservation holds in O',

$$0 = x^{\bar{\alpha}'}{}_{,\bar{\beta}}\left(P^{\bar{\beta}}(I) - P^{\bar{\beta}}(F)\right) = P^{\bar{\alpha}'}(I) - P^{\bar{\alpha}'}(F).$$

It is necessary to write physical law as a tensor equation, as in Eq. (2.33). Then, if the law holds for one observer, it holds for all. Such a decay is always handled in SR, hence the bars over momentum indexes. The earth is freely falling in the metric due mainly to the sun. As will be seen, it is an inertial system. In the weak gravity of an earth laboratory, there is no noticeable earthly gravitational effect on these particles. Though the decay is handled in SR, the general principle holds for all frames.

Other vectors are easy to define. For example, one could define a force vector $F^{\mu} \equiv \frac{dP^{\mu}}{d\tau}$. The force components on each of two, at rest, charged particles can be calculated. After a Lorentz transform, to a frame in which

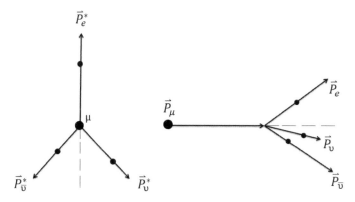

Fig. 2.6   Muon decay: On the left is the view in the muon's rest frame (O). On the right is the view in the laboratory, frame O'.

the particles are moving, one notes magnetic effects enter the scene. In this way, the laws of electromagnetism can be derived from Coulomb's law and SR. Such material is useful for an electrodynamics course, but not pertinent for further study of GR.

Problems

1. In an inertial frame of SR and in rectangular coordinates, the metric for the spatial coordinates is η_{ij}. Obtain the metric and determinant of the metric in spherical coordinates. The relationships between coordinates are as follows:

$$x = r \sin\theta \cos\phi,$$
$$y = r \sin\theta \sin\phi,$$
$$z = r \cos\theta,$$
$$r = [(x)^2 + (y)^2 + (z)^2]^{1/2},$$
$$\phi = \tan^{-1} y/x,$$
$$\theta = \cos^{-1}(z/[(x)^2 + (y)^2 + (z)^2]^{1/2}).$$

2. Show in general that $x^{\mu'}{}_{,\nu''} = g_{\nu''\xi''}g^{\mu'\chi'}x^{\xi''}{}_{,\chi'}$. Then, using the information in Problem 1, show explicitly that, in an inertial frame where $x^{\bar\mu'}$, $x^{\bar\nu''}$ are cylindrical, spherical coordinates,

$$x^{\bar\mu'}{}_{,\bar\nu''} = g_{\bar\nu''\bar\xi''}g^{\bar\mu'\bar\chi'}x^{\bar\xi''}{}_{,\bar\chi'}.$$

3. Show that Eq. (2.6) leads to Eq. (2.7) for the Lorentz transform, discussed in Chapter 1. Show explicitly for this transform that

$$x^{\bar\mu}{}_{,\bar\nu'} = \eta^{\mu\alpha}\eta_{\nu'\beta'}x^{\bar\beta'}{}_{,\bar\alpha}.$$

4. Prove that the 4-volume element $[-\det(g_{\mu\nu})]^{1/2}dx^0 dx^1 dx^2 dx^3$ is an invariant. Use the following information. In a rectangular coordinate frame, in flat space,

$$d^4V = [-\det(\eta_{\mu\nu})]^{1/2}dx^{\bar 0}dx^{\bar 1}dx^{\bar 2}dx^{\bar 3} = dx^{\bar 0}dx^{\bar 1}dx^{\bar 2}dx^{\bar 3}.$$

In another frame, calculus yields

$$d^4V = J\,dx^0 dx^1 dx^2 dx^3, \quad J \equiv \text{Jacobian},$$
$$J = \frac{\partial(x^{\bar 0}x^{\bar 1}x^{\bar 2}x^{\bar 3})}{\partial(x^0 x^1 x^2 x^3)},$$
$$= \det(\Delta),$$

where Δ is the following matrix:

$$
\Delta = \begin{pmatrix}
x^{\bar{0}}{}_{,0} & x^{\bar{0}}{}_{,1} & x^{\bar{0}}{}_{,2} & x^{\bar{0}}{}_{,3} \\
x^{\bar{1}}{}_{,0} & x^{\bar{1}}{}_{,1} & x^{\bar{1}}{}_{,2} & x^{\bar{1}}{}_{,3} \\
x^{\bar{2}}{}_{,0} & x^{\bar{2}}{}_{,1} & x^{\bar{2}}{}_{,2} & x^{\bar{2}}{}_{,3} \\
x^{\bar{3}}{}_{,0} & x^{\bar{3}}{}_{,1} & x^{\bar{3}}{}_{,2} & x^{\bar{3}}{}_{,3}
\end{pmatrix}.
$$

Now relate $g_{\mu\nu}$ to Δ.

5. Show from the transform equations that $g_{\mu\nu}T^{\nu}{}_{\beta}$, where $T^{\nu}{}_{\beta}$ is a mixed tensor of rank 2, is a covariant tensor of rank 2. If $\frac{dg_{\alpha\beta}}{d\tau} = 0$, show $U_{\mu}\frac{dU^{\mu}}{d\tau} = 0$, where U^{μ} is the velocity.

6. Show from the transform equations that $\delta^{\mu}{}_{\nu}$ is a mixed tensor of rank 2. What are $g^{\mu\nu}g_{\xi\chi}$, $g^{\mu\mu}g_{\nu\nu}$, $g^{\mu\nu}g_{\mu\nu}$, $g^{\mu\mu}g_{\mu\mu}$ and $g^{\mu\nu}V_{\xi}$, where V_{ξ} is a vector?

7. In the twin problem, the rocket moved relative to frame O with spatial components of velocity and acceleration in the z-direction only, and O′ was at rest in the rocket.

   (a) Suppose now that the rocket was an inertial frame, and O′ had nonzero spatial components of 3-velocity and 3-acceleration, in all directions relative to the rocket. Here $v^{\bar{i}'} = \frac{dx^{\bar{i}'}}{dt'}$, and $a^{\bar{i}'} = \frac{dv^{\bar{i}'}}{dt'}$. If the rocket moves with speed V in the z-direction relative to O, what are $a^{\bar{j}}$ in terms of V, $v^{\bar{i}'}$, and $a^{\bar{i}'}$? You already know $v^{\bar{j}}$ in terms of V and $v^{\bar{i}'}$.

   (b) Use Eqs. (2.21) and (2.22) to evaluate the invariant $(\eta_{\mu\nu}A^{\mu}A^{\nu})^{1/2}$. Convince yourself that if one inertial observer sees an object accelerating, all inertial observers see it accelerating.

8. In the twin problem, what is the velocity of the rocket with respect to the inertial twin, the time on the inertial twin's clock, and the distance of the rocket from the inertial twin, as a function of the time on the rocket clock? Check that your answers give the expected results at $\tau = T$, $2T$, $3T$, $4T$. Let the rocket clock time be $0 \le \tau_i \le T$ for each of the four legs. Similarly, let V_i, t_i, z_i be the variables to find for each of the legs. For t_{tot} and z_{tot}, one has to add contributions from more than one leg.

9. A person on the surface of earth lives 80 y as reckoned by his/her proper-time clock. What would the lifetime be in a space station at a distance $2.5R_e$ from earth's center, on his/her proper-time clock? What would the earth clock read at time of death in the space station? Repeat the above calculation for lifetime in a mine, at a distance $0.99R_e$ from

earth's center, assuming a spherically symmetric earth with uniform density. In order to simplify things, neglect the velocity effect.

10. Use the data for the Pound–Rebka experiment. For fixed source and detector, what is the difference in energy in eV between photons emitted at the bottom and photons arriving at the top of the tower.

11. A laser emits light of frequency ν in the $-z$-direction. The laser is mounted to a land rocket, that also moves in the $-z$-direction with speed V with respect to the earth. A second such rocket has a mirror mounted on it and reflects the light received from the laser. It moves with speed V' relative to the earth. What frequency is measured at the laser rocket for the reflected light. Consider all possible cases: $V > V'$, $V < V'$ and V' moving in the $\pm z$-direction relative to earth. The visible spectrum lies between $(0.43\text{--}0.75) \times 10^{13}\,\mathrm{Hz}$. Suppose, $V = 500\,\mathrm{mph} \ll c$, what is the magnitude of V' such that mirror rocket travel in the $\mp z$-directions yields reflected light at the laser, with the lowest and highest visible frequencies.

12. The most efficient rocket one can conceive of would convert its rest mass to photons. The photons would part from the remaining mass in one direction and propel the remaining mass in the reverse direction. What fraction of the original mass remains when its speed is 0.9, 0.925, 0.95, 0.975?

13. At present high-energy proton accelerators new particles and their anti-particles can be produced by the reactions,

$$p + p \to X + \bar{X} + p + p, \quad M_X \gg M_p.$$

There are two types of accelerators: one beam and a fixed target, liquid hydrogen, or colliding beams with equal and opposite momenta. In accelerating protons to the highest energies, it is the kinetic energy of the beam(s) that determines cost. Calculate the minimum kinetic energy to produce the final state for both types of accelerators and determine which is cheaper to construct? In the case of Higgs boson production, $M_X = 0.125\,\mathrm{TeV}/c^2$. All known conservation laws would be satisfied without also requiring the production of the anti-Higgs. In this case, what is the minimum collider kinetic energy? Could the Higgs be produced at any current fixed target facility?

14. A particle of rest mass M decays into two particles of rest masses M_1 and M_2. If M is traveling with momentum $P\hat{e}_z$ in the laboratory, calculate the cosine of the opening angle $\cos\theta_o$ between the two particles, in the laboratory. Make the calculation in terms of the cosine of the

emission angle $\cos\theta^*$, with respect to $\hat{e}_z$, of M_1 in the rest frame of M. Graph this function for $M = 1$ m, $M_{1,2} = 0.3$, 0.5 m and $P = 1$, 1.5 m. Note that there is a maximum opening angle between the decay particles in the laboratory that decreases as P increases.

15. In Chapter 9, cosmology is dealt with. There, the cosmic microwave background (CMB) radiation has a Planck distribution with a temperature $T_0 = 2.73\,\text{K}$. In the early universe, the temperature was much higher, and the energy, of the photons $E_\gamma = h\nu = k_B T$ was much greater. Here, k_B is Boltzmann's constant and T is the absolute temperature. When the energy was high enough particle, anti-particle pairs could be produced via,

$$\gamma + \gamma \to X + \bar{X}.$$

What is the minimum photon energy required to produce electron–positron $(X = e)$ pairs? The scale factor of the universe Q, compared to what it is now Q_0, is $Q/Q_0 = T_0/T$. What was this ratio when $e^\pm$, pairs could no longer be produced because the universe was expanding and cooling?

16. The reaction, considered in problem 15, explains why an upper limit for gamma-ray energy in cosmic rays is observed. Very energetic gamma rays travel long distances to reach earth. They are traveling through a sea of 2.73 K photons, as they started out long ago. They are quite likely to collide with one and produce electron–positron pairs, before reaching us. Calculate the observable maximum gamma-ray energy in MeV. There is evidence for a diffuse X-ray background with energy $\approx 2 \times 10^6 k_B T_0$ that would lower the above observable maximum energy considerably. Recalculate that energy for this diffuse X-ray background.

17. The Robertson–Walker metric describes the development of the universe. Its elements are: $g_{00} = -1$, $g_{33} = g_{rr} = [Q(t)]^2$, $g_{22} = g_{\phi\phi} = (Q(t)r\sin\theta)^2$, $g_{11} = g_{\theta\theta} = (Q(t)r)^2$ and $g_{\mu\nu} = 0, (\mu \neq \nu)$. Here, r is the radial coordinate from an origin, say the center of the sun. Consider a galaxy at radial coordinate R. The galaxy can be considered to be on the surface of a sphere of radial coordinate R. What is the circumference and area of the sphere? What is the proper distance from the origin? As $Q(t)$ is getting bigger with time, one sees that the proper distance between any origin and a radial coordinate is expanding. Space is being created between the two. Experiment indicates that there is a non-understood "dark energy" driving the expansion, and that it will not be reversed by the gravity of the existing mass.

18. On the scale of the solar system, the universal expansion is much too small to be observed. The metric that describes the earth's orbit about the sun is the static Schwarzschild metric. Here, $g_{00} = -(g_{rr})^{-1} = -(1 - 2M_s/r)$, $g_{\theta\theta} = r^2$, $g_{\phi\phi} = \sin^2\theta g_{\theta\theta}$, where M_s is the sun's mass and r is the earth's radial coordinate. What is the value of L_p/r correct to 0.1%? Be careful and creative; you cannot use this metric in the sun's interior.

19. Show that even in a complicated metric, like the Schwarzschild metric described in Problem 18, an observer still measures unity for the speed of light.

Chapter 3

Covariant Differentiation, Equations of Motion

3.1 Differentiation of Invariants and Vectors

In the previous chapters, the importance of tensors in spacetime was stressed. Any time a new quantity is encountered, it will have to be checked to see if it is a tensor. If it isn't, its transformation properties are not obvious. Construction of new tensors has, so far, taken the form of products of known tensors or total differentiation with respect to τ. For example, $(d\tau)^2 = dr_\mu dr^\mu$, $g^{\mu\nu}g_{\xi\nu} = \delta^\mu_{\ \xi} = \delta_\xi^{\ \mu}$, or $U^\mu = \frac{dr^\mu}{d\tau}$. From studies of the calculus of 3-vectors, one recalls that partial differentiation with respect to the coordinates produces new 3-vectors and scalars through the gradient and divergence operations. In spacetime, such partial differentiation also leads to important new tensors.

Consider an invariant that is a function of position, $\Phi = \Phi(x^\mu) = \Phi(x^{\mu'})$, e.g., $d\tau$. It has no index associated with it. Taking the partial derivative with respect to a coordinate yields

$$\Phi_{,\nu} = x^{\xi'}_{\ ,\nu}\, \Phi_{,\xi'}. \tag{3.1}$$

However, this is the rule for the transformation of a covariant vector and so another vector is added to our arsenal.

The gradient of a scalar Φ is given by $g^{\mu\nu}\Phi_{,\nu}$ because in an inertial frame the expected results for the spatial components are obtained

$$\bar{\nabla}^{\bar\mu}\Phi \equiv g^{\bar\mu\bar\nu}\Phi_{,\bar\nu} = \eta^{\mu\nu}\Phi_{,\bar\nu},$$
$$\vec{\nabla}\Phi = \Phi_{,x}\,\hat{e}_x + \Phi_{,y}\,\hat{e}_y + \Phi_{,z}\,\hat{e}_z. \tag{3.2}$$

41

If cylindrical coordinates were chosen, the gradient also has the expected form

$$\nabla^\rho \Phi = g^{\rho\rho}\Phi,_\rho = \Phi,_\rho, \quad \nabla^\phi \Phi = g^{\phi\phi}\Phi,_\phi = \Phi,_\phi / \rho^2,$$

$$\vec{\nabla}\Phi - \Phi,_z \hat{e}_z = \Phi,_\rho \vec{e}_\rho + (\Phi,_\phi / \rho^2)\vec{e}_\phi = \Phi,_\rho \hat{e}_\rho + (\Phi,_\phi / \rho)\hat{e}_\phi. \tag{3.3}$$

The case of spherical coordinates is left as an exercise.

A similar partial differentiation of a vector can be performed. For example, for the position vector in an inertial frame in rectangular coordinates,

$$dr^{\bar\mu} = dx^{\bar\mu}, \quad \text{thus } \partial r^{\bar\mu} = \partial x^{\bar\mu},$$

$$r^{\bar\mu},_{\bar\nu} = x^{\bar\mu},_{\bar\nu} = \delta^{\bar\mu}_{\bar\nu}.$$

So the partial derivative of the component of a vector with respect to a coordinate appears to be an element of a tensor of rank 2, see Problem 2.6. However, in cylindrical or spherical coordinates, none of the $r^{\bar\mu'}$ depend on coordinate $\phi = x^{\bar2'}$, so that $r^{\bar2'},_{\bar2'} = 0$. In general, terms in addition to the partial derivative of a vector with respect to a coordinate, are needed to obtain an object that transforms as a tensor of rank 2.

In order to see how to handle this, return to the calculus of 3-vectors. For the position vector in the plane perpendicular to $\hat{e}_z$, in cylindrical coordinates, and using the results of the rotation indicated in Fig. 1.1,

$$\vec{r} = r^{\bar i'}\vec{e}_{\bar i'} = r^{\bar1'}\vec{e}_{\bar1'} = r^\rho \vec{e}_\rho = \rho\vec{e}_\rho,$$

$$\vec{r},_{\bar1'} = \vec{r},_\rho = \rho,_\rho \vec{e}_\rho + \rho\vec{e}_\rho,_\rho = \vec{e}_\rho = \hat{e}_\rho = \vec{e}_{\bar1'},$$

$$\vec{r},_{\bar2'} = \vec{r},_\phi = \rho,_\phi \vec{e}_\rho + \rho\vec{e}_\rho,_\phi = \rho\hat{e}_\rho,_\phi = \rho(\cos\phi\hat{e}_x + \sin\phi\hat{e}_y),_\phi \tag{3.4}$$

$$= \rho(-\sin\phi\hat{e}_x + \cos\phi\hat{e}_y) = \rho\hat{e}_\phi = \vec{e}_\phi = \vec{e}_{\bar2'}, \quad \text{thus,}$$

$$(\vec{r},_{\bar j'})^{\bar i'} = \delta_{\bar j'}^{\bar i'}.$$

This result is the same as the rectangular coordinate result.

In tensor notation, the terms needed, in addition to the partial derivative, lead to the covariant derivative,

$$V^\mu;_\nu \equiv V^\mu,_\nu + \Gamma^\mu_{\beta\nu}V^\beta. \tag{3.5}$$

Note the use of the semi colon to indicate covariant differentiation. The left-hand side is the $\binom{\mu}{\nu}$ element of the mixed tensor that is the covariant derivative of vector V. The quantity $\Gamma^{\mu}_{\beta\nu}$ is called the Christoffel (C) symbol, and it can be calculated solely from the $g_{\mu\nu}$. From above, the first term is all that is needed in $x^{\bar{\mu}}$ coordinates. There $\Gamma^{\bar{\mu}}_{\bar{\beta}\bar{\nu}} = 0$. This derivative obeys the same algebra, when applied to products, as does the partial derivative because it does so in $x^{\bar{\mu}}$ coordinates. There, it is just the partial derivative. The other terms take into account the way the unit vectors, or in tensor notation, the way the metric tensor depends on the coordinates.

None of the terms on the right-hand side of Eq. (3.5) transforms like a tensor of rank 2 but the sum does. For an arbitrary vector V, a straightforward, but lengthy proof, using the tensor transformation rule, is left as Problem 2. Then, it follows that

$$\delta^{\bar{\mu}}_{\bar{\nu}} = r^{\bar{\mu}}{}_{,\bar{\nu}} = r^{\bar{\mu}}{}_{;\bar{\nu}},$$

$$\delta^{\alpha}_{\beta} = r^{\alpha}{}_{;\beta}. \tag{3.6}$$

The first line of the above equation establishes a tensor equation so it holds for any coordinates, yielding the second line.

In order to illustrate that the covariant derivative yields expected results, consider cylindrical coordinates of an inertial frame. As shall be seen, the only nonzero C symbols are $\Gamma^{\bar{2}'}_{\bar{1}'\bar{2}'} = \Gamma^{\bar{2}'}_{\bar{2}'\bar{1}'} = \Gamma^{\phi}_{\rho\phi} = \Gamma^{\phi}_{\phi\rho} = 1/\rho$ and $\Gamma^{\bar{1}'}_{\bar{2}'\bar{2}'} = \Gamma^{\rho}_{\phi\phi} = -\rho$. The covariant derivative of the position vector in the plane perpendicular to $\hat{e}_z$ is as follows:

$$r^{\bar{i}'}{}_{;\bar{j}'} = r^{\bar{i}'}{}_{,\bar{j}'} + \Gamma^{\bar{i}'}_{\bar{j}'\bar{\alpha}'} r^{\bar{\alpha}'}, \text{ with components,}$$

$$r^{\rho}{}_{;\rho} = r^{\rho}{}_{,\rho} + \Gamma^{\rho}_{\rho\bar{\alpha}'} r^{\bar{\alpha}'} = \rho{}_{,\rho} = 1,$$

$$r^{\rho}{}_{;\phi} = r^{\rho}{}_{,\phi} + \Gamma^{\rho}_{\phi\bar{\alpha}'} r^{\bar{\alpha}'} = \Gamma^{\rho}_{\phi\phi} r^{\phi} = 0,$$

$$r^{\phi}{}_{;\phi} = r^{\phi}{}_{,\phi} + \Gamma^{\phi}_{\phi\bar{\alpha}'} r^{\bar{\alpha}'} = \Gamma^{\phi}_{\phi\rho} r^{\rho} = \rho^{-1}\rho = 1, \tag{3.7}$$

$$r^{\phi}{}_{;\rho} = r^{\phi}{}_{,\rho} + \Gamma^{\phi}_{\rho\bar{\alpha}'} r^{\bar{\alpha}'} = \Gamma^{\phi}_{\rho\phi} r^{\phi} = 0,$$

$$r^{\bar{i}'}{}_{;\bar{j}'} = \delta^{\bar{i}'}_{\bar{j}'}.$$

Note that Eq. (3.7) agrees with Eq. (3.6) and the 3-vector results of Eq. (3.4).

The scalar divergence of a vector is defined Div $V = V^{\mu}{}_{;\mu}$ because in an inertial frame the expected results are obtained for the spatial part. This is shown first in rectangular, and then in cylindrical coordinates for the

(ρ, ϕ) terms,

$$V^{\bar{\mu}}{}_{,\bar{\mu}} = V^t{}_{,t} + V^x{}_{,x} + V^y{}_{,y} + V^z{}_{,z}.$$

$$V^{\rho}{}_{;\rho} = V^{\rho}{}_{,\rho} + \Gamma^{\rho}_{\bar{\alpha}'\rho} V^{\bar{\alpha}'} = V^{\rho}{}_{,\rho},$$

$$V^{\phi}{}_{;\phi} = V^{\phi}{}_{,\phi} + \Gamma^{\phi}_{\bar{\alpha}'\phi} V^{\bar{\alpha}'} = V^{\phi}{}_{,\phi} + \Gamma^{\phi}_{\rho\phi} V^{\rho} = V^{\phi}{}_{,\phi} + V^{\rho}/\rho,$$

$$V^{\rho}{}_{;\rho} + V^{\phi}{}_{;\phi} = V^{\rho}{}_{,\rho} + V^{\rho}/\rho + V^{\phi}{}_{,\phi} = (\rho V^{\rho}){}_{,\rho}/\rho + V^{\phi}{}_{,\phi}. \tag{3.8}$$

If the vector is the gradient of a scalar Φ, then in an inertial frame, the expected wave equation for Φ is obtained. As above, first in rectangular and then in cylindrical coordinates. Using Eqs. (3.2) and (3.3),

$$(\nabla^{\bar{\mu}}\Phi);_{\bar{\mu}} = -\Phi_{,t\ ,t} + \Phi_{,x\ ,x} + \Phi_{,y\ ,y} + \Phi_{,z\ ,z}.$$

$$(\nabla^{\rho}\Phi);_{\rho} + (\nabla^{\phi}\Phi);_{\phi} = (\nabla^{\rho}\Phi)_{,\rho} + \Gamma^{\rho}_{\bar{\alpha}'\rho}\nabla^{\bar{\alpha}'}\Phi + (\nabla^{\phi}\Phi)_{,\phi} + \Gamma^{\phi}_{\bar{\alpha}'\phi}\nabla^{\bar{\alpha}'}\Phi$$

$$= (\nabla^{\rho}\Phi)_{,\rho} + \Gamma^{\phi}_{\rho\phi}\nabla^{\rho}\Phi + (\nabla^{\phi}\Phi)_{,\phi}$$

$$= \Phi_{,\rho\ ,\rho} + \Phi_{,\rho}/\rho + \Phi_{,\phi\ ,\phi}/\rho^2$$

$$= (\rho\Phi_{,\rho})_{,\rho}/\rho + \Phi_{,\phi\ ,\phi}/\rho^2. \tag{3.9}$$

The covariant derivative of a covariant vector can be calculated by considering an invariant, such as the scalar product of two vectors $V_\mu U^\mu$. Using Eq. (3.1) and renaming summed over indexes, yields

$$V_{\mu;\nu} U^\mu + V_\mu U^\mu{}_{;\nu} = (V_\mu U^\mu);_\nu = (V_\mu U^\mu)_{,\nu},$$

$$V_{\mu;\nu} U^\mu + V_\mu(U^\mu{}_{,\nu} + \Gamma^\mu_{\alpha\nu}U^\alpha) = V_{\mu,\nu} U^\mu + V_\mu U^\mu{}_{,\nu},$$

$$V_{\mu;\nu} U^\mu = V_{\mu,\nu} U^\mu - V_\mu \Gamma^\mu_{\alpha\nu} U^\alpha \tag{3.10}$$

$$= V_{\mu,\nu} U^\mu - V_\alpha \Gamma^\alpha_{\mu\nu} U^\mu,$$

$$V_{\mu;\nu} = V_{\mu,\nu} - \Gamma^\alpha_{\mu\nu} V_\alpha.$$

The C symbol can be shown to be symmetric in its covariant indexes because the order of partial differentiation doesn't matter,

$$\Phi;_{\bar{\mu}};_{\bar{\nu}} = \Phi_{,\bar{\mu}\ ,\bar{\nu}} = \Phi_{,\bar{\nu}\ ,\bar{\mu}} = \Phi;_{\bar{\nu}};_{\bar{\mu}}. \tag{3.11}$$

This is a tensor equation and so it holds in general, thus,

$$\Phi_{,\mu};_\nu = \Phi;_\mu;_\nu = \Phi;_\nu;_\mu = \Phi_{,\nu};_\mu,$$

$$\Phi_{,\mu\ ,\nu} - \Gamma^\alpha_{\mu\nu}\Phi_{,\alpha} = \Phi_{,\nu\ ,\mu} - \Gamma^\alpha_{\nu\mu}\Phi_{,\alpha}, \tag{3.12}$$

$$\Gamma^\alpha_{\nu\mu} = \Gamma^\alpha_{\mu\nu}.$$

Due to this property $\Gamma^{\alpha}_{\mu\nu}$ has a term like $x^{\bar{\beta}}{}_{,\mu}{}_{,\nu}$. This term satisfies Eq. (3.12) and in the inertial frame $x^{\bar{\beta}}{}_{,\mu}{}_{,\nu} \rightarrow x^{\bar{\beta}}{}_{,\bar{\mu}}{}_{,\bar{\nu}} = 0$, as required. To take care of the contravariant index and get rid of the index $\bar{\beta}$, one could try

$$\Gamma^{\alpha}_{\mu\nu} = x^{\alpha}{}_{,\bar{\beta}}\, x^{\bar{\beta}}{}_{,\mu}{}_{,\nu}. \tag{3.13}$$

It turns out that Eq. (3.13) holds. However, a more useful form, see below, in terms of the metric tensor and its derivatives, will be used. The proof of Eq. (3.13) starts with the more useful form and is left as Problem 3.

3.2 Differentiation of Tensors

Given two vectors V and W, the product, $V^{\mu}W_{\nu}$, transforms like a mixed tensor of rank 2, and its covariant derivative yields

$$
\begin{aligned}
T^{\mu}{}_{\nu;\alpha} = (V^{\mu}W_{\nu})_{;\alpha} &= V^{\mu}{}_{;\alpha}\,W_{\nu} + V^{\mu}W_{\nu;\alpha} \\
&= (V^{\mu}{}_{,\alpha} + \Gamma^{\mu}_{\beta\alpha}V^{\beta})W_{\nu} + V^{\mu}(W_{\nu,\alpha} - \Gamma^{\beta}_{\nu\alpha}W_{\beta}) \\
&= (V^{\mu}W_{\nu})_{,\alpha} + \Gamma^{\mu}_{\beta\alpha}V^{\beta}W_{\nu} - \Gamma^{\beta}_{\nu\alpha}V^{\mu}W_{\beta} \\
&= T^{\mu}{}_{\nu,\alpha} + \Gamma^{\mu}_{\beta\alpha}T^{\beta}{}_{\nu} - \Gamma^{\beta}_{\nu\alpha}T^{\mu}{}_{\beta},
\end{aligned} \tag{3.14}
$$

yielding a mixed tensor of rank 3. The contravariant index requires a positive sign, while the covariant index requires a negative sign for the C symbol. In a similar manner, one obtains the covariant derivatives of a covariant or contravariant tensor of rank 2. If the rank is higher, say n, then n C symbols with appropriate signs are needed. In the case of the metric tensor,

$$g^{\mu\nu}{}_{;\alpha} = g^{\mu\nu}{}_{,\alpha} + \Gamma^{\mu}_{\beta\alpha}g^{\beta\nu} + \Gamma^{\nu}_{\alpha\beta}g^{\mu\beta} = 0, \tag{3.15}$$

$$g_{\mu\nu;\alpha} = g_{\mu\nu,\alpha} - \Gamma^{\beta}_{\mu\alpha}g_{\beta\nu} - \Gamma^{\beta}_{\alpha\nu}g_{\mu\beta} = 0. \tag{3.16}$$

The reason the above tensors are zero is that in an inertial frame $g_{\bar{\mu}\bar{\nu};\bar{\alpha}} = \eta_{\mu\nu;\bar{\alpha}} = \eta_{\mu\nu,\bar{\alpha}} = 0$. As this is a tensor equation, it holds in all frames, and leads to the more useful form for $\Gamma^{\lambda}_{\mu\nu}$,

$$
\begin{aligned}
0 = g_{\mu\nu;\alpha} &+ g_{\mu\alpha;\nu} - g_{\alpha\nu;\mu} \\
&= g_{\mu\nu,\alpha} + g_{\mu\alpha,\nu} - g_{\alpha\nu,\mu} - \Gamma^{\beta}_{\mu\alpha}g_{\beta\nu} - \Gamma^{\beta}_{\alpha\nu}g_{\mu\beta} \\
&\quad - \Gamma^{\beta}_{\mu\nu}g_{\beta\alpha} - \Gamma^{\beta}_{\alpha\nu}g_{\mu\beta} + \Gamma^{\beta}_{\mu\alpha}g_{\beta\nu} + \Gamma^{\beta}_{\mu\nu}g_{\alpha\beta}, \\
2g_{\mu\beta}\Gamma^{\beta}_{\alpha\nu} &= (g_{\mu\nu,\alpha} + g_{\mu\alpha,\nu} - g_{\alpha\nu,\mu}), \\
2g^{\mu\lambda}g_{\mu\beta}\Gamma^{\beta}_{\alpha\nu} &= 2\delta^{\lambda}_{\beta}\Gamma^{\beta}_{\alpha\nu} = g^{\mu\lambda}(g_{\mu\nu,\alpha} + g_{\mu\alpha,\nu} - g_{\alpha\nu,\mu}), \\
\Gamma^{\lambda}_{\alpha\nu} &= g^{\mu\lambda}(g_{\mu\nu,\alpha} + g_{\mu\alpha,\nu} - g_{\alpha\nu,\mu})/2.
\end{aligned} \tag{3.17}
$$

Once the metric tensor is obtained, the C symbols are easily calculated. For example, in cylindrical coordinates $x^{\mu'}$, the previous quoted results for the nonzero C symbols are substantiated. Let $(\bar{i}', \bar{j}', \bar{k}')$ take on the values $(\bar{1}', \bar{2}')$. Note that nonzero metric elements are $g_{\bar{1}'\bar{1}'} = 1$ and $g_{\bar{2}'\bar{2}'} = \rho^2$,

$$\Gamma^{\bar{i}'}_{\bar{j}'\bar{k}'} - g^{\bar{i}'\bar{m}'}(g_{\bar{j}'\bar{m}',\bar{k}'} \mid g_{\bar{k}'\bar{m}',\bar{j}'} - g_{\bar{j}'\bar{k}',\bar{m}'})/2$$

$$= g^{\bar{i}'\bar{i}'}(g_{\bar{i}'\bar{j}',\bar{k}'} + g_{\bar{i}'\bar{k}',\bar{j}'} - g_{\bar{j}'\bar{k}',\bar{i}'})/2$$

$$= g^{\bar{i}'\bar{i}'}(\delta^{\bar{i}'}_{\bar{j}'}g_{\bar{i}'\bar{i}',\bar{k}'} + \delta^{\bar{i}'}_{\bar{k}'}g_{\bar{i}'\bar{i}',\bar{j}'} - \delta^{\bar{j}'}_{\bar{k}'}g_{\bar{k}'\bar{k}',\bar{i}'})/2,$$

$$\Gamma^{\bar{1}'}_{\bar{1}'\bar{1}'} = g^{\bar{1}'\bar{1}'}g_{\bar{1}'\bar{1}',\bar{1}'}/2 = 0, \quad \Gamma^{\bar{2}'}_{\bar{2}'\bar{2}'} = g^{\bar{2}'\bar{2}'}g_{\bar{2}'\bar{2}',\bar{2}'}/2 = 0,$$

$$\Gamma^{\bar{1}'}_{\bar{1}'\bar{2}'} = g^{\bar{1}'\bar{1}'}g_{\bar{1}'\bar{1}',\bar{2}'}/2 = 0, \quad \Gamma^{\bar{2}'}_{\bar{1}'\bar{1}'} = -g^{\bar{2}'\bar{2}'}g_{\bar{1}'\bar{1}',\bar{2}'}/2 = 0,$$

$$\Gamma^{\bar{2}'}_{\bar{1}'\bar{2}'} = \Gamma^{\phi}_{\rho\phi} = g^{\bar{2}'\bar{2}'}g_{\bar{2}'\bar{2}',\bar{1}'}/2 = \rho^{-2}\rho^2{}_{,\rho}/2 = 1/\rho,$$

$$\Gamma^{\bar{1}'}_{\bar{2}'\bar{2}'} = \Gamma^{\rho}_{\phi\phi} = -g^{\bar{1}'\bar{1}'}g_{\bar{2}'\bar{2}',\bar{1}'}/2 = -\rho^2{}_{,\rho}/2 = -\rho.$$

3.3   Gravity and the Locally Inertial Frame

Liberal use is made of the thought experiments proposed by Einstein, illustrated in Figs. 3.1 and 3.2, to crystallize his ideas concerning GR. When gravity is included, it acts everywhere. At first it isn't clear that an inertial frame can be found. Standing in an elevator, gravity is experienced by the floor pushing up on our feet. There is no upward acceleration as gravity counters this force. If the elevator started accelerating upward, a stronger upward push would be experienced. However, an elevator observer couldn't tell whether the effect was due to a stronger gravitational force, or a force due to a machine capable of lifting the elevator.

If the elevator cable broke and the elevator observer released some objects from rest, a camera fixed to the elevator would show all objects remaining at rest with respect to each other. It wouldn't matter whether some objects were more or less massive or if they were made of different materials. Also the elevator observer would no longer feel an upward force from the floor. A camera fixed to the earth would show all objects falling with the same acceleration. This is because inertial mass and gravitational mass are the same. For example, from Newton's law of gravity,

$$F = M_I a = M_G M/d^2, \quad a = M/d^2.$$

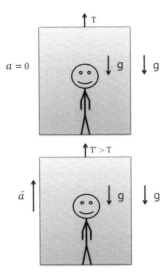

Fig. 3.1 Top, observer at rest in an elevator is also rest with respect to earth. Bottom, elevator and observer accelerating upward with respect to earth. The accelerating observer feels a stronger upward force from the floor and could conclude gravity is stronger than when at rest.

Fig. 3.2 Freely falling elevator observer releases various objects and notes they do not fall; concluding that there is no gravity. Earth observer, not freely falling, notes all objects in the elevator are falling, with the same acceleration because of gravity.

Here M is the attracting mass, M_I is the inertial mass of the accelerating object, M_G is the gravitational mass of the accelerating object, and d is the distance between the masses.

As Einstein put it, the elevator observer is free falling and doesn't experience gravity. The results of all non-gravity experiments in the elevator will be the same as the results obtained in an inertial frame. This is the "weak" principle of equivalence. The "strong" principle says the results of any experiment will be the same as those in an inertial frame. It is somewhat ironic; gravity is a complicated subject, but its distinguishing feature is, if you are freely falling in uniform gravity, it is not experienced.

Gravitational strength varies with position. One would experience gravity through tidal forces. However, in an arbitrarily small volume a free falling observer will, to first order, not experience gravity. For example, no one is conscious of freely falling due to the gravitational influence of the sun. This is the essence of a locally inertial frame. In this arbitrarily small frame, coordinates $x^{\bar{\mu}}$ can be found with origin at arbitrary point P, such that,

$$g_{\bar{\alpha}\bar{\beta}}(x^{\bar{\mu}}) = \eta_{\alpha\beta} + O(x^{\bar{\mu}})^2,$$

$$g_{\bar{\alpha}\bar{\beta}}|_P = \eta_{\alpha\beta}, \tag{3.18}$$

$$g_{\bar{\alpha}\bar{\beta},\bar{\chi}}|_P = g_{\bar{\alpha}\bar{\beta};\bar{\chi}}|_P = 0, \tag{3.19}$$

$$g_{\bar{\alpha}\bar{\beta},\bar{\chi},\bar{\xi}}|_P \neq 0. \tag{3.20}$$

Another way to see the locally inertial frame in our lives is easy. People live on the curved surface of an approximately spherical earth. However, a person occupies a very small portion of this surface, so that the person is not conscious of the curvature. As will be seen, gravity causes space curvature.

3.4   Local Flatness Theorem

The existence of a locally inertial frame means that curved space has a flat space tangent at any point. Above, it was noted that at any point P in spacetime, one can find a rectangular coordinate system $x^{\bar{\mu}}$, such that at P the metric is the metric of SR. Then, its first partial derivative vanishes, but not necessarily its second partial. That is, the metric near P is approximately that of SR, the differences being second order in the coordinates.

The proof begins by noting that there is some relation between an arbitrary coordinate system and the locally inertial system, $x^\mu = x^\mu(x^{\bar\nu})$, and all orders of the partial derivatives, $x^\mu{}_{,\bar\nu}$,..., exist. The latter and the metric are expanded in a Taylor series, shown below to second order. The expansion is about P, for coordinates close to P,

$$x^\mu{}_{,\bar\alpha} = x^\mu{}_{,\bar\alpha}|_P + x^\mu{}_{,\bar\alpha}{}_{,\bar\chi}|_P \Delta^{\bar\chi} + x^\mu{}_{,\bar\alpha}{}_{,\bar\chi}{}_{,\bar\xi}|_P \Delta^{\bar\chi}\Delta^{\bar\xi}/2,$$

$$x^\mu{}_{,\bar\alpha} x^\nu{}_{,\bar\beta}\, g_{\mu\nu} = g_{\bar\alpha\bar\beta}$$

$$= g_{\bar\alpha\bar\beta}|_P + g_{\bar\alpha\bar\beta}{}_{,\bar\chi}|_P \Delta^{\bar\chi} + g_{\bar\alpha\bar\beta}{}_{,\bar\chi}{}_{,\bar\xi}|_P \Delta^{\bar\chi}\Delta^{\bar\xi}/2, \qquad (3.21)$$

$$\Delta^{\bar\chi} = x^{\bar\chi} - x^{\bar\chi}|_P, \text{ thus, } \Delta^{\bar\chi}|_P = 0,$$

$$\Delta^{\bar\chi}{}_{,\bar\lambda}|_P = \delta^{\bar\chi}{}_{\bar\lambda}, \; (\Delta^{\bar\chi}\Delta^{\bar\xi})_{,\bar\lambda}{}_{,\bar\gamma}|_P = \delta^{\bar\chi}{}_{\bar\lambda}\delta^{\bar\xi}{}_{\bar\gamma} + \delta^{\bar\chi}{}_{\bar\gamma}\delta^{\bar\xi}{}_{\bar\lambda}.$$

At P, the differences in the coordinates $\Delta^{\bar\chi}$ vanish, and Eq. (3.21) becomes

$$g_{\bar\alpha\bar\beta}|_P = [x^\mu{}_{,\bar\alpha} x^\nu{}_{,\bar\beta}\, g_{\mu\nu}]|_P = \eta_{\alpha\beta}.$$

This equality is a set of ten independent equations because the metric tensor $g_{\mu\nu}$ is symmetric and has ten independent values. However, there are 16 independent first-order partial derivatives available $x^\mu{}_{,\bar\alpha}$, more than enough to satisfy these equations.

If Eq. (3.21) is partially differentiated with respect to $x^{\bar\lambda}$ and evaluated at P, the leading term, second order, and all the higher order terms, with remaining coordinate differences vanish. This yields

$$g_{\bar\alpha\bar\beta}{}_{,\bar\lambda}|_P = [(x^\mu{}_{,\bar\alpha} x^\nu{}_{,\bar\beta}\, g_{\mu\nu})_{,\bar\lambda}]|_P$$

$$= [x^\mu{}_{,\bar\alpha} x^\nu{}_{,\bar\beta}\, g_{\mu\nu,\bar\lambda} + x^\mu{}_{,\bar\alpha}{}_{,\bar\lambda} x^\nu{}_{,\bar\beta}\, g_{\mu\nu} + x^\mu{}_{,\bar\alpha} x^\nu{}_{,\bar\beta}{}_{,\bar\lambda}\, g_{\mu\nu}]|_P = 0.$$

This equality is a set of 40 independent equations because the metric tensor is symmetric and there are 40 independent values of $g_{\mu\nu,\bar\lambda}$, ten for each $\bar\lambda$. As the first-order partial derivatives have been used, these equations can be satisfied, if there are at least 40 independent second-order partial derivatives, $x^\mu{}_{,\bar\alpha}{}_{,\bar\lambda}$ and $x^\nu{}_{,\bar\beta}{}_{,\bar\lambda}$. Since the order of partial differentiation is unimportant, there are exactly 40 such derivatives, ten for each μ or ν, just enough to satisfy the equations.

If Eq. (3.21) is partially differentiated twice with respect to, $x^{\bar\lambda}$ and $x^{\bar\gamma}$, and evaluated at P, the leading term, first-order term, and all the terms higher than second order, with remaining coordinate differences vanish. Then,

$$g_{\bar{\alpha}\bar{\beta},\bar{\lambda},\bar{\gamma}}\,|_P = [(x^{\mu}{}_{,\bar{\alpha}}\,x^{\nu}{}_{,\bar{\beta}}\,g_{\mu\nu})_{,\bar{\lambda},\bar{\gamma}}]|_P, \equiv Q + R + S,$$

$$Q = [x^{\mu}{}_{,\bar{\alpha}}\,x^{\nu}{}_{,\bar{\beta}}\,g_{\mu\nu,\bar{\gamma},\bar{\lambda}}]|_P,$$

$$R = [(x^{\mu}{}_{,\bar{\alpha},\bar{\lambda}}\,x^{\nu}{}_{,\bar{\beta}} + x^{\mu}{}_{,\bar{\alpha}}\,x^{\nu}{}_{,\bar{\beta},\bar{\lambda}})g_{\mu\nu,\bar{\gamma}}$$

$$+ (x^{\mu}{}_{,\bar{\alpha},\bar{\gamma}}\,x^{\nu}{}_{,\bar{\beta}} + x^{\mu}{}_{,\bar{\alpha}}\,x^{\nu}{}_{,\bar{\beta},\bar{\gamma}})g_{\mu\nu,\bar{\lambda}}]|_P,$$

$$S = [(x^{\mu}{}_{,\bar{\alpha},\bar{\gamma},\bar{\lambda}}\,x^{\nu}{}_{,\bar{\beta}} + x^{\mu}{}_{,\bar{\alpha},\bar{\gamma}}\,x^{\nu}{}_{,\bar{\beta},\bar{\lambda}} + x^{\mu}{}_{,\bar{\alpha},\bar{\lambda}}\,x^{\nu}{}_{,\bar{\beta},\bar{\gamma}}$$

$$+ x^{\mu}{}_{,\bar{\alpha}}\,x^{\nu}{}_{,\bar{\beta},\bar{\gamma},\bar{\lambda}})g_{\mu\nu}]|_P.$$

The equality $Q + R + S$ is a set of 100 independent equations because there are 100 (10×10) independent values of $g_{\mu\nu,\bar{\gamma},\bar{\lambda}}$. The metric tensor is symmetric and the order of partial differentiation is unimportant. As the first- and second-order partial derivatives have been used, at least 100 independent third-order partial derivatives $x^{\mu}{}_{,\bar{\alpha},\bar{\gamma},\bar{\lambda}}$, are required for a solution. However, as the order of partial differentiation is unimportant there are only 80 such derivatives, 20 for each μ, $\bar{\alpha}\bar{\gamma}\bar{\lambda} = 000, 001, 002, 003, 011, 012,$ 013, 022, 023, 033, 111, 112, 113, 122, 123, 133, 222, 223, 233, 333. Thus, there aren't enough parameters to satisfy the 100 equations. The 20 lacking parameters are needed, as shall be seen, to describe the curvature of space.

At P, the locally inertial rectangular coordinates $x^{\bar{\mu}}$ may be very complicated functions of the rectangular coordinates of a non-inertial observer. The metric tensor elements in these coordinates are $\eta_{\alpha\beta}$ and the C symbols are zero, but derivatives of the C symbols don't, in general, vanish. Transforming to other frames, even if rectangular coordinates are used, will lead to non-constant metric tensors and C symbols. However, the covariant derivative of the metric tensor remains $g_{\mu\nu;\xi} = 0$, and the result for the C symbols in terms of the metric tensor and its derivatives Eq. (3.17) remains valid.

3.5   GR Equations of Motion

In the locally inertial frame specified by $x^{\bar{\mu}}$, all objects travel in a straight line without acceleration. The equation of motion of a particle with rest mass is as follows:

$$\frac{d^2 x^{\bar{\mu}}}{d\tau^2} = \frac{dU^{\bar{\mu}}}{d\tau} = 0. \tag{3.22}$$

Using, Eq. (3.13), the above, in another frame becomes

$$0 = \frac{dU^{\bar{\alpha}}}{d\tau} = \frac{d(x^{\bar{\alpha}}{}_{,\mu} U^{\mu})}{d\tau} = x^{\bar{\alpha}}{}_{,\mu} \frac{dU^{\mu}}{d\tau} + U^{\mu} \frac{d}{d\tau} x^{\bar{\alpha}}{}_{,\mu}$$

$$= x^{\bar{\alpha}}{}_{,\mu} \frac{dU^{\mu}}{d\tau} + U^{\mu} x^{\bar{\alpha}}{}_{,\mu,\nu} \frac{dx^{\nu}}{d\tau}$$

$$= x^{\bar{\alpha}}{}_{,\mu} \frac{dU^{\mu}}{d\tau} + U^{\mu} x^{\bar{\alpha}}{}_{,\mu,\nu} U^{\nu}$$

$$= x^{\beta}{}_{,\bar{\alpha}} \left(x^{\bar{\alpha}}{}_{,\mu} \frac{dU^{\mu}}{d\tau} + U^{\mu} x^{\bar{\alpha}}{}_{,\mu,\nu} U^{\nu} \right) = \delta^{\beta}{}_{\mu} \frac{dU^{\mu}}{d\tau} + U^{\mu} U^{\nu} \Gamma^{\beta}{}_{\mu\nu}$$

$$= \frac{dU^{\beta}}{d\tau} + U^{\mu} U^{\nu} \Gamma^{\beta}{}_{\mu\nu}. \tag{3.23}$$

The equation of motion involves the C symbols, which depend on the metric tensor.

Since a photon has $d\tau = 0$, the above cannot be used as its equation of motion. One must substitute another parameter, say dq. Allowable parameters are called affine parameters. For example, $d\tau = kdq$ yields an allowable dq, where for photons $k = 0$. The parameter q describes the path, such that as the photon moves $\frac{dx^{\mu}}{dq} = W^{\mu}$ is the tangent vector, with the property,

$$W_{\mu} W^{\mu} = g_{\mu\nu} W^{\mu} W^{\nu} = g_{\mu\nu} \frac{dx^{\mu}}{dq} \frac{dx^{\nu}}{dq} = \frac{g_{\mu\nu} dx^{\mu} dx^{\nu}}{dqdq}$$

$$= \left(\frac{d\tau}{dq} \right)^{2} = 0. \tag{3.24}$$

In the inertial frame, the geodesic is a straight line and the tangent vector doesn't change. The result in another frame is, following the procedure after Eq. (3.22),

$$0 = \frac{dW^{\bar{\alpha}}}{dq} = \frac{d(x^{\bar{\alpha}}{}_{,\mu} W^{\mu})}{dq} = x^{\bar{\alpha}}{}_{,\mu,\nu} W^{\mu} W^{\nu} + x^{\bar{\alpha}}{}_{,\mu} \frac{dW^{\mu}}{dq}$$

$$= x^{\beta}{}_{,\bar{\alpha}} \left(x^{\bar{\alpha}}{}_{,\mu,\nu} W^{\mu} W^{\nu} + x^{\bar{\alpha}}{}_{,\mu} \frac{dW^{\mu}}{dq} \right)$$

$$= \Gamma^{\beta}{}_{\mu\nu} W^{\mu} W^{\nu} + \frac{dW^{\beta}}{dq}$$

$$= \frac{d^{2} x^{\beta}}{dq^{2}} + \Gamma^{\beta}{}_{\mu\nu} \frac{dx^{\mu}}{dq} \frac{dx^{\nu}}{dq}. \tag{3.25}$$

Problems

1. What is the covariant derivative $\delta^\mu{}_{\beta;\nu}$ in a frame where gravity must be taken into account? Evaluate the partial derivative $\delta^\mu{}_{\beta,\nu}$ and show $x^\beta{}_{,\alpha'}\, x^\mu{}_{,\rho'}\, x^{\rho'}{}_{,\nu}{}_{,\beta} = -x^\mu{}_{,\alpha'}{}_{,\nu}$.

2. Show that, in general, the partial derivative of a contravariant vector $V^\mu{}_{,\nu}$ doesn't transform like a tensor. Show that the C symbol $\Gamma^\mu_{\nu\xi}$ doesn't transform like a tensor. Show that the covariant derivative of a vector $V^\mu{}_{;\nu}$ transforms like a mixed tensor of rank 2.

3. Derive the alternate form for the C symbol $\Gamma^\xi_{\mu\nu} = x^\xi{}_{,\bar\alpha}\, x^{\bar\alpha}{}_{,\mu}{}_{,\nu}$ from $\Gamma^\xi_{\mu\nu} = g^{\xi\beta}(g_{\mu\beta,\nu} + g_{\nu\beta,\mu} - g_{\mu\nu,\beta})/2$.

4. Harmonic coordinates satisfy $g^{\mu\nu}\Gamma^\xi_{\mu\nu} = 0$. Show from the transform of the C symbol in Problem 2, that if a set of coordinates is not harmonic one can always solve a second-order differential equation and get to a harmonic set. Find that equation. Specifically, show that for inertial systems, rectangular coordinates are harmonic, but cylindrical and spherical coordinates are not.

5. The importance of writing the laws of physics as tensor equations has been discussed. However, the laws of electromagnetism, conservation of charge and Maxwell's equations, are usually written in 3-vector notation. In vacuum and in an SR frame with rectangular coordinates, these equations are in naturalized units:

$$\vec\nabla \cdot \vec J + \rho_{,t} = 0,$$

$$\vec\nabla \cdot \vec B = 0, \qquad \vec B = \vec\nabla \times \vec A,$$

$$\vec\nabla \times \vec E + \vec B_{,t} = 0, \qquad \vec E = -\vec\nabla\Psi - \vec A_{,t},$$

$$\vec\nabla \cdot \vec E - \rho/\epsilon_0 = 0 = \vec\nabla \cdot \vec E - \mu_0\rho,$$

$$\vec\nabla \times \vec B - \vec E_{,t} = \mu_0\vec J,$$

where the vector $J^\mu = (\rho, \vec J)$ has as components the (charge, current) densities, $(\vec E, \vec B)$ are the (electric, magnetic) fields, the vector $A^\mu = (\Psi, \vec A)$ has as components the (scalar, vector) potentials, and (ϵ_0, μ_0) are the (permittivity, permeability) of free space.

Show that the divergence of the vector J^μ leads to charge conservation as a tensor equation. Show that the components of the 3-vectors

$(\vec{E}, \vec{B})$ can be written as elements of a tensor of rank 2,

$$F_{\nu\mu} = A_{\mu,\nu} - A_{\nu,\mu} = -F_{\mu\nu}, \text{ where}$$

$$A_{\mu} = (-\Psi, \vec{A}),$$

and that the four Maxwell equations can be written as two tensor equations,

$$F^{\mu\nu}{}_{;\nu} = \mu_0 J^{\mu} = J^{\mu}/\epsilon_0,$$

$$0 \equiv F_{\mu\nu;\xi} + F_{\xi\mu;\nu} + F_{\nu\xi;\mu}.$$

6. Consider the following useful metric, written in spherical coordinates, $x^{0,1,2,3} = t, \theta, \phi, r$,

$$(d\tau)^2 = -g_{\mu\nu}dx^{\mu}dx^{\nu}$$
$$= \exp[2\Phi(r)](dt)^2 - (\exp[2\Delta(r)](dr)^2 + (r)^2[(d\theta)^2 + (\sin\theta d\phi)^2].$$

With the correct forms for $\Phi(r)$ and $\Delta(r)$ this is either the metric of an inertial frame in spherical coordinates or the Schwarzschild metric used to discuss the motion of light and planets due to the sun's gravity. Find all the nonzero C symbols. If confined to the surface of a sphere, so that, $r = a = $ constant, one would be aware of curvature, even in SR. In this case, what are the nonzero C symbols?

7. Show explicitly for the metric of Problem 6 that $g_{\mu\nu;\chi} = 0$. Now specialize to an inertial frame. In this frame what are the nonzero C symbols? You have already found the result if $r = a$.

8. For the metric of Problem 6 in an inertial frame, use the C symbols calculated in Problem 7 to find the gradient and Laplacian of a scalar function of position.

9. What does the quantity $V^{\mu}{}_{;\lambda}{}_{;\gamma}$ transform like? In the presence of gravity,

$$V^{\mu}{}_{;\lambda}{}_{;\gamma} - V^{\mu}{}_{;\gamma}{}_{;\lambda} = -R^{\mu}{}_{\beta\lambda\gamma}V^{\beta}.$$

Evaluate $R^{\mu}{}_{\beta\lambda\gamma}$ in terms of the C symbols and their partial derivatives. Then, argue that

$$V_{\mu;\lambda}{}_{;\gamma} - V_{\mu;\gamma}{}_{;\lambda} = R^{\beta}{}_{\mu\lambda\gamma}V_{\beta}.$$

Show $R^{\mu}{}_{\beta\lambda\gamma} = -R^{\mu}{}_{\beta\gamma\lambda}$. What does the quantity $R^{\mu}{}_{\beta\lambda\gamma}$ transform like? This quantity is the Riemann curvature tensor. It plays an important role in GR and will be discussed in much more detail in later chapters.

10. If $T^{\mu_1\mu_2\cdots}{}_{\nu_1\nu_2\ldots}$ is a general mixed tensor, use the results of Problem 9 to evaluate,

$$T' = T^{\mu_1\mu_2\cdots}{}_{\nu_1\nu_2\ldots;\lambda\;;\gamma} - T^{\mu_1\mu_2\cdots}{}_{\nu_1\nu_2\ldots;\gamma\;;\lambda},$$

where T' is a tensor of the same rank as that of $T + 2$. Apply this result to the metric tensor, in an arbitrary frame, to give an alternate proof that $R_{\mu\nu\lambda\gamma} = -R_{\nu\mu\lambda\gamma}$.

11. A point particle of finite rest mass moves in a gravity free region of empty space. What are the Newtonian equations of motion in cylindrical coordinates, (t, ρ, ϕ, z)? Show that the GR equations of motion give the same results.

12. For the metric of Problem 6, find the GR equations of motion for a particle with rest mass? Show that a planar solution is possible. In this case, find another constant of the motion.

Chapter 4

Curvature

4.1 Geodesics

The thought experiments of Einstein, illustrated in Fig. 4.1, show that in the presence of gravity light moves in a curved path. The top of the figure shows two equivalent observers O, one in gravity free space, and the other freely falling in a region of uniform gravity. They observe that a horizontally traveling light ray enters and exits their capsule a distance L above the floor. The bottom of the figure shows an observer O', not freely falling due to gravity. O' also observes, that for the freely falling capsule, light entered and exited the same distance above the floor. However, according to O' the exit point will have fallen in the time light crossed the capsule. Thus, the light also must have fallen or moved in a downward-curved path. The conclusion to be drawn is that gravity affects light, a break from Newtonian physics.

In the geometry of flat space, geodesics are the paths of minimum distance between two points, for motion with constant velocity, or the paths that minimize the travel time. Light in empty space certainly fits this case and before GR the phrase, "light travels in straight lines," was often heard. This is because gravity is so weak, that the deviation from a straight line path, was too small to be observed. GR knowledgeable observers where gravity acts, but not freely falling, know that nothing can make the trip between two points faster than light. Thus, the "straight lines" or geodesics, are actually curved paths.

Mathematically, the GR equations of motion also lead to geodesics for the motion of a particle with rest mass. Making use of Eqs. (3.16), (3.17)

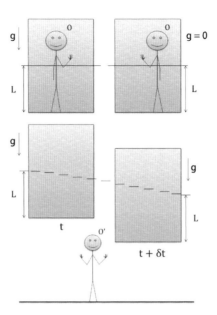

Fig. 4.1 Top: equivalent observers, one in gravity free space, the other freely falling in a region of uniform gravity. They both observe horizontally traveling light enter and leave their capsule a distance L above the floor. Bottom: an observer, not freely falling due to gravity, observes light travels in a curved trajectory, to enter and leave the same distance above the floor.

and (3.23), and renaming summed over indexes when necessary yields,

$$
\begin{aligned}
\frac{d(g_{\mu\nu}U^\mu U^\nu)}{d\tau} &= \frac{dg_{\mu\nu}}{d\tau}U^\mu U^\nu + g_{\mu\nu}\left(\frac{dU^\mu}{d\tau}U^\nu + U^\mu\frac{dU^\nu}{d\tau}\right) \\
&= g_{\mu\nu,\gamma}\frac{dx^\gamma}{d\tau}U^\mu U^\nu + g_{\mu\nu}\left(\frac{dU^\mu}{d\tau}U^\nu + U^\mu\frac{dU^\nu}{d\tau}\right) \\
&= g_{\mu\nu,\gamma}U^\gamma U^\mu U^\nu - g_{\mu\nu}(\Gamma^\mu_{\alpha\beta}U^\nu + U^\mu\Gamma^\nu_{\alpha\beta})U^\alpha U^\beta \\
&= (g_{\alpha\beta,\gamma} - g_{\mu\gamma}\Gamma^\mu_{\alpha\beta} - g_{\gamma\nu}\Gamma^\nu_{\alpha\beta})U^\alpha U^\beta U^\gamma \\
&= (g_{\lambda\beta}\Gamma^\lambda_{\gamma\alpha} + g_{\alpha\lambda}\Gamma^\lambda_{\beta\gamma} - g_{\mu\gamma}\Gamma^\mu_{\alpha\beta} - g_{\gamma\nu}\Gamma^\nu_{\alpha\beta})U^\alpha U^\beta U^\gamma \\
&= (g_{\mu\gamma}\Gamma^\mu_{\beta\alpha} + g_{\gamma\nu}\Gamma^\nu_{\beta\alpha} - g_{\mu\gamma}\Gamma^\mu_{\alpha\beta} - g_{\gamma\nu}\Gamma^\nu_{\alpha\beta})U^\alpha U^\beta U^\gamma = 0,
\end{aligned}
$$
$$
g_{\mu\nu}U^\mu U^\nu = K = -1.
$$

This is true as the initial condition is $(d\tau)^2 = -g_{\mu\nu}dx^\mu dx^\nu$. Then $-1 = g_{\mu\nu}U^\mu U^\nu$ and the expression for $(d\tau)^2$ always holds along the path.

One can introduce a parameter q to describe the path of a body under the influence of gravity. Given the value of that parameter, the position on the path is determined. However, algebra allows manipulations so that the final result is independent of q. The proper time elapsed when the body moves from A to B is

$$T_{BA} = \int_A^B \frac{d\tau}{dq} dq = \int_A^B dq \frac{(-g_{\mu\nu} dx^\mu dx^\nu)^{1/2}}{dq} = \int_A^B dq \left(-g_{\mu\nu} \frac{dx^\mu}{dq} \frac{dx^\nu}{dq}\right)^{1/2}.$$

Vary the path from $x^\mu(q)$ to $x^\mu(q) + \delta x^\mu(q)$ while keeping the endpoints fixed. The change in the elapsed time is as follows:

$$\delta T_{BA} = -\int_A^B dq \frac{N}{D}, \quad \text{where}$$

$$N = \left(g_{\mu\nu,\beta} \, \delta x^\beta \frac{dx^\mu}{dq} \frac{dx^\nu}{dq} + g_{\mu\nu} \frac{d(\delta x^\mu)}{dq} \frac{dx^\nu}{dq} + g_{\mu\nu} \frac{dx^\mu}{dq} \frac{d(\delta x^\nu)}{dq}\right),$$

$$D = 2 \left(-g_{\mu\nu} \frac{dx^\mu}{dq} \frac{dx^\nu}{dq}\right)^{1/2} = \frac{2d\tau}{dq},$$

$$\delta T_{BA} = -\int_A^B \frac{dqdq}{2d\tau} \left(g_{\mu\nu,\beta} \, \delta x^\beta \frac{dx^\mu}{dq} \frac{dx^\nu}{dq} + g_{\mu\nu} \frac{d(\delta x^\mu)}{dq} \frac{dx^\nu}{dq}\right.$$

$$\left. + g_{\mu\nu} \frac{dx^\mu}{dq} \frac{d(\delta x^\nu)}{dq}\right)$$

$$= -\int_A^B \frac{dqdqd\tau}{d\tau d\tau} \left(g_{\mu\nu,\beta} \, \delta x^\beta \frac{dx^\mu}{dq} \frac{dx^\nu}{dq} + 2g_{\mu\nu} \frac{d(\delta x^\mu)}{dq} \frac{dx^\nu}{dq}\right) / 2$$

$$= -\int_A^B d\tau \left(g_{\mu\nu,\beta} \, \delta x^\beta U^\mu U^\nu / 2 + g_{\mu\nu} \frac{d(\delta x^\mu)}{d\tau} U^\nu\right).$$

The second term $T2$ can be integrated by parts with the proviso that $\delta x^\mu = 0$ at the endpoints. After such integration, use of the expression for the C symbol simplifies the result,

$$T2 = \int_A^B d\tau [g_{\mu\nu} U^\nu] \frac{d(\delta x^\mu)}{d\tau} = \int_A^B [g_{\mu\nu} U^\nu] d(\delta x^\mu) \equiv \int_A^B [W] d(V)$$

$$= (WV)_{\text{endpoints}} - \int_A^B V dW = -\int_A^B V dW = -\int_A^B d\tau V \frac{dW}{d\tau}$$

$$= -\int_A^B d\tau \delta x^\mu \frac{d(g_{\mu\nu} U^\nu)}{d\tau}$$

$$= - \int_A^B d\tau \delta x^\mu \left(g_{\mu\nu} \frac{dU^\nu}{d\tau} + U^\nu U^\lambda g_{\mu\nu,\lambda} \right)$$

$$= - \int_A^B d\tau \delta x^\beta \left(g_{\beta\nu} \frac{dU^\nu}{d\tau} + U^\mu U^\lambda g_{\beta\mu,\lambda} \right).$$

This yields

$$\delta T_{BA} = - \int_A^B d\tau \delta x^\beta F,$$

$$F = g_{\mu\lambda,\beta} U^\mu U^\lambda /2 - U^\mu U^\lambda g_{\beta\mu,\lambda} - g_{\beta\nu} \frac{dU^\nu}{d\tau}$$

$$= g_{\mu\lambda,\beta} U^\mu U^\lambda /2 - \left(U^\mu U^\lambda g_{\beta\mu,\lambda} + U^\mu U^\lambda g_{\beta\mu,\lambda} \right)/2 - g_{\beta\nu} \frac{dU^\nu}{d\tau}$$

$$= g_{\mu\lambda,\beta} U^\mu U^\lambda /2 - \left(U^\mu U^\lambda g_{\beta\mu,\lambda} + U^\lambda U^\mu g_{\beta\lambda,\mu} \right)/2 - g_{\beta\nu} \frac{dU^\nu}{d\tau}$$

$$= U^\mu U^\lambda (g_{\mu\lambda,\beta} - g_{\beta\mu,\lambda} - g_{\beta\lambda,\mu})/2 - g_{\beta\nu} \frac{dU^\nu}{d\tau}, \text{ but}$$

$$\Gamma^\nu_{\mu\lambda} = -g^{\alpha\nu}(g_{\mu\lambda,\alpha} - g_{\alpha\mu,\lambda} - g_{\alpha\lambda,\mu})/2,$$

$$g_{\beta\nu}\Gamma^\nu_{\mu\lambda} = -\delta^\alpha_\beta(g_{\mu\lambda,\alpha} - g_{\alpha\mu,\lambda} - g_{\alpha\lambda,\mu})/2 = -(g_{\mu\lambda,\beta} - g_{\beta\mu,\lambda} - g_{\beta\lambda,\mu})/2,$$

$$F = -g_{\beta\nu} \left[\Gamma^\nu_{\mu\lambda} U^\mu U^\lambda + \frac{dU^\nu}{d\tau} \right] = 0, \quad \delta T_{BA} = 0. \tag{4.1}$$

The last line follows from the equations of motion. Thus, the time for the trip is an extremum, usually a minimum, so the path is a geodesic. As the object moves from point to point, the clock attached to the object ticks at the rate appropriate for the gravitational strength at the point. For a time extremum, a straight line path is unlikely. The object, like light, moves in a curved path, so that it takes an extremum of proper time to make the trip.

However, for light the element of proper time always vanishes $d\tau = 0$. Thus, another explanation for the geodesic that light must travel along is required. The one deduced by Einstein is that gravity curves space and all objects, even photons, must travel along that curvature. It's similar to us traveling on the surface of a spherical earth. Empty space or vacuum is no longer seen to be just volume where objects can position themselves. It is rather like a fabric, that can be pulled and contorted by the gravitational influence of faraway objects.

On a cosmological scale, things are even more complicated. Space is always being created between any two points. This is due to an unknown form of "dark energy," that is dependent on the size of the universe. In Chapter 9, devoted to cosmology, it will be seen that this is now the dominant energy because the universe has expanded to such a large volume. It's a really strange universe we inhabit, and just common sense, obtained from every day observation, could never lead you to its inner workings.

4.2 Parallel Transport

The characterization of curvature starts with the concept of parallel transport. On a flat surface, as in Fig. 4.2, draw an arbitrary closed path $ABCA$. Here, a circle is used, so that at various places along the path, some tangent vectors $\vec{W}$, that can point in all directions, are shown. For parallel transport start at A, draw on the surface, a small parallel transport vector $\vec{V}$ in any direction. Proceed to a neighboring point on the path, draw on the surface a small transport vector, as parallel as possible to the $\vec{V}$ previously drawn. On a flat surface, it is possible to draw the vector exactly parallel. When once again at A, the identical vectors would be redrawn. In this sense, a flat surface has no intrinsic curvature. A cylinder can be constructed by rolling a flat sheet, and so has no intrinsic curvature.

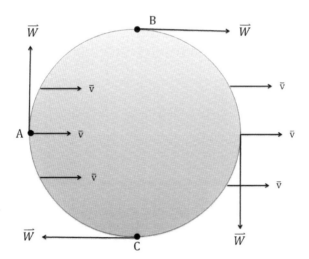

Fig. 4.2 An arbitrary closed curve, here a circle, on a flat surface, with tangent vectors $\vec{W}$. Parallel transport vectors $\vec{V}$, at any point on the curve, can be drawn on the surface parallel to each other.

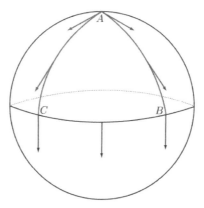

Fig. 4.3   A sphere has intrinsic curvature, so that there are paths for which the parallel transport vectors do not line up, after traversing a closed curve on the surface.

A sphere cannot be made from a flat sheet. It has intrinsic curvature. One can find at least one path on the sphere's surface, as in Fig. 4.3, for which the vectors $\vec{V}$, would not repeat. Pick the path $ABCA$, such that B and C are on the equator, and A is at a pole. At A, start with a vector $\vec{V}$ on the sphere's surface, that is tangent to an arc of longitude. As one proceeds to B, along the longitude, a new parallel transport vector cannot be drawn on the surface, exactly parallel to $\vec{V}$. The best one can do is draw that vector along the tangent vector. At B that vector is perpendicular to the equator, and remains so as one proceeds to C. From there the return to A is again along a longitude. The parallel transport vectors on the surface will be opposite the tangent vectors. Upon reaching A, the final parallel transport vector is different from the initial one.

In spacetime, these vectors have four components V^μ, W^μ. At any point P, one can go to a locally inertial frame. In a small enough neighborhood of P, as you proceed along the curve specified by affine parameter q and $W^{\bar{\nu}} = \frac{dx^{\bar{\nu}}}{dq}$, the vector $V^{\bar{\mu}}$ is constant. This leads to a tensor equation, that is taken as the frame invariant definition of parallel transport of V^μ along W^ν,

$$0 = \left.\frac{dV^{\bar{\mu}}}{dq}\right|_P = V^{\bar{\mu}}{}_{,\bar{\nu}}\frac{dx^{\bar{\nu}}}{dq} = W^{\bar{\nu}}V^{\bar{\mu}}{}_{,\bar{\nu}} = W^{\bar{\nu}}V^{\bar{\mu}}{}_{;\bar{\nu}} = W^{\nu}V^{\mu}{}_{;\nu}. \quad (4.2)$$

The last equality occurs because the next to last equality established the above as a tensor equation.

In flat space, the geodesics are straight lines. These are the only curves that parallel transport their tangent vector. In curved space, the geodesics are drawn as "straight" as possible, by demanding parallel transport of the tangent vector. This leads to the equation of motion in terms of q. From Eq. (4.2), with $V^{\bar{\nu}} = W^{\bar{\nu}}$,

$$0 = W^\nu W^\mu{}_{;\nu} = W^\nu(W^\mu{}_{,\nu} + W^\beta \Gamma^\mu_{\beta\nu}) = W^\nu W^\mu{}_{,\nu} + W^\nu W^\beta \Gamma^\mu_{\beta\nu}$$

$$= \frac{dx^\nu}{dq}\left(\frac{dx^\mu}{dq}\right)_{,\nu} + \frac{dx^\nu}{dq}\frac{dx^\beta}{dq}\Gamma^\mu_{\beta\nu} = \frac{d^2 x^\mu}{dq^2} + \frac{dx^\nu}{dq}\frac{dx^\beta}{dq}\Gamma^\mu_{\beta\nu}. \qquad (4.3)$$

The reader should note that this equation is just Eq. (3.25).

The above equations can be used to quantify curvature. Consider travel along the elemental closed curve $ABCDA$, that bounds area on a spherical surface, such as that blown up in Fig. 4.4. The curvature is not noticeable for such a small area. The angles, $(\theta, \phi) = (x^1, x^2)$, vary, such that along element AB, $x^1 = a$, x^2 varies; along BC, $x^2 = b + \delta b$, x^1 varies; along CD, $x^1 = a + \delta a$, x^2 varies, and along AD, $x^2 = b$, x^1 varies. Vector V^μ defined at A is parallel transported around the curve. The curve is given, which specifies W^i, and vector V^μ is arbitrary, so that for each path

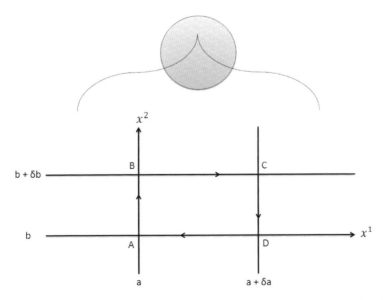

Fig. 4.4   Blowup of an elemental area on a spherical surface that is bound by closed curve $ABCDA$.

element,

$$0 = V^\mu{}_{;i}, \quad V^\mu{}_{,i} = -V^\beta \Gamma^\mu_{i\beta}, \quad i = 1, 2. \tag{4.4}$$

For example, the first trip element yields $V^\mu(AB) = \int_A^B dx^2 V^\mu{}_{,2} |_{x^1=a}$. Then,

$$V^\mu(AB) = -\int_A^B dx^2 (V^\beta \Gamma^\mu_{2\beta})|_{x^1=a} = -\int_b^{b+\delta b} dx^2 (V^\beta \Gamma^\mu_{2\beta})|_{x^1=a},$$

$$V^\mu(BC) = -\int_B^C dx^1 (V^\beta \Gamma^\mu_{1\beta})|_{x^2=b+\delta b} = -\int_a^{a+\delta a} dx^1 (V^\beta \Gamma^\mu_{1\beta})|_{x^2=b+\delta b},$$

$$V^\mu(CD) = -\int_C^D dx^2 (V^\beta \Gamma^\mu_{2\beta})|_{x^1=a+\delta a} = -\int_{b+\delta b}^b dx^2 (V^\beta \Gamma^\mu_{2\beta})|_{x^1=a+\delta a},$$

$$V^\mu(DA) = -\int_D^A dx^1 (V^\beta \Gamma^\mu_{1\beta})|_{x^2=b} = -\int_{a+\delta a}^a dx^1 (V^\beta \Gamma^\mu_{1\beta})|_{x^2=b}.$$

The minus sign can be used to flip the limits in any of the above integrals.

Add the element contributions, make use of the definition of the partial derivative, and find that the change of the vector when once again at A is

$$\delta V^\mu(A) = \int_b^{b+\delta b} dx^2 [(V^\beta \Gamma^\mu_{2\beta})|_{x^1=a+\delta a} - (V^\beta \Gamma^\mu_{2\beta})|_{x^1=a}]$$

$$- \int_a^{a+\delta a} dx^1 [(V^\beta \Gamma^\mu_{1\beta})|_{x^2=b+\delta b} - (V^\beta \Gamma^\mu_{1\beta})|_{x^2=b}]$$

$$= -\left[\int_a^{a+\delta a} \delta b\, dx^1 (V^\beta \Gamma^\mu_{1\beta})_{,2} - \int_b^{b+\delta b} \delta a\, dx^2 (V^\beta \Gamma^\mu_{2\beta})_{,1} \right]$$

$$= -\delta a \delta b [(V^\beta \Gamma^\mu_{1\beta})_{,2} - (V^\beta \Gamma^\mu_{2\beta})_{,1}]$$

$$= -\delta a \delta b (V^\beta{}_{,2}\, \Gamma^\mu_{1\beta} + V^\beta \Gamma^\mu_{1\beta,2} - V^\beta{}_{,1}\, \Gamma^\mu_{2\beta} - V^\beta \Gamma^\mu_{2\beta,1})$$

$$= -\delta a \delta b (-V^\nu \Gamma^\beta_{2\nu}\, \Gamma^\mu_{1\beta} + V^\beta \Gamma^\mu_{1\beta,2} + V^\nu \Gamma^\beta_{1\nu}\, \Gamma^\mu_{2\beta} - V^\beta \Gamma^\mu_{2\beta,1})$$

$$= \delta a \delta b V^\beta (\Gamma^\nu_{2\beta} \Gamma^\mu_{1\nu} - \Gamma^\mu_{1\beta,2} - \Gamma^\nu_{1\beta} \Gamma^\mu_{2\nu} + \Gamma^\mu_{2\beta,1}).$$

The last line results from using Eq. (4.4) to rewrite the partial derivatives, $V^\beta{}_{,(1,2)}$. The quantity in parentheses is nonzero for this curve because a sphere has intrinsic curvature. In Problem 3, one calculates, in the absence of gravity, a finite curvature on a spherical surface, and none on a cylindrical surface.

4.3   Curvature Tensors

When gravity is present and there are no boundary surfaces, there is no reason to allow just $x^{1,2}$ to vary, so let them be replaced by generalized coordinates, $x^{\gamma,\lambda}$. Then, the above quantity in parentheses is defined as the Riemann curvature tensor,

$$R^{\mu}_{\beta\gamma\lambda} = \Gamma^{\nu}_{\lambda\beta}\Gamma^{\mu}_{\gamma\nu} - \Gamma^{\mu}_{\gamma\beta,\lambda} - \Gamma^{\nu}_{\gamma\beta}\Gamma^{\mu}_{\lambda\nu} + \Gamma^{\mu}_{\lambda\beta,\gamma}. \tag{4.5}$$

The proof that $R^{\mu}_{\beta\gamma\lambda}$ is a tensor was carried out in Problem 3.9, where the following results were obtained for vector V^{μ}:

$$V^{\mu}_{;\lambda\,;\gamma} - V^{\mu}_{;\gamma\,;\lambda} = R^{\mu}_{\beta\lambda\gamma}V^{\beta}, \quad R^{\mu}_{\beta\lambda\gamma} = -R^{\mu}_{\beta\gamma\lambda}. \tag{4.6}$$

Since the covariant derivative of a tensor is a tensor, the left-hand side of Eq. (4.6) is a tensor of rank 3. The right-hand side of the first equality must be a tensor. Since V^{β} is a tensor of rank 1, $R^{\mu}_{\beta\lambda\gamma}$ must be a tensor of rank 4.

This tensor simplifies for rectangular coordinates in a locally inertial frame because the C symbols, but not their partial derivatives vanish, and $g^{\bar\mu\bar\nu}_{;\bar\chi} = g^{\bar\mu\bar\nu}_{,\bar\chi} = \eta^{\mu\nu}_{,\bar\chi} = 0$,

$$\Gamma^{\bar\mu}_{\bar\lambda\bar\beta,\bar\gamma} = [(g^{\bar\alpha\bar\mu}[g_{\bar\beta\bar\alpha,\bar\lambda} + g_{\bar\lambda\bar\alpha,\bar\beta} - g_{\bar\beta\bar\lambda,\bar\alpha}]),_{\bar\gamma}]/2$$

$$= (g^{\bar\alpha\bar\mu}_{,\bar\gamma}[g_{\bar\beta\bar\alpha,\bar\lambda} + g_{\bar\lambda\bar\alpha,\bar\beta} - g_{\bar\beta\bar\lambda,\bar\alpha}] + g^{\bar\alpha\bar\mu}[g_{\bar\beta\bar\alpha,\bar\lambda} + g_{\bar\lambda\bar\alpha,\bar\beta} - g_{\bar\beta\bar\lambda,\bar\alpha}],_{\bar\gamma})/2$$

$$= (g^{\bar\alpha\bar\mu}[g_{\bar\beta\bar\alpha,\bar\lambda} + g_{\bar\lambda\bar\alpha,\bar\beta} - g_{\bar\beta\bar\lambda,\bar\alpha}],_{\bar\gamma})/2$$

$$= g^{\bar\alpha\bar\mu}(g_{\bar\beta\bar\alpha,\bar\lambda},_{\bar\gamma} + g_{\bar\lambda\bar\alpha,\bar\beta},_{\bar\gamma} - g_{\bar\beta\bar\lambda,\bar\alpha},_{\bar\gamma})/2.$$

Similarly, noting the metric tensor is symmetric, and the order of partial differentiation is immaterial,

$$\Gamma^{\bar\mu}_{\bar\gamma\bar\beta,\bar\lambda} = g^{\bar\alpha\bar\mu}(g_{\bar\beta\bar\alpha,\bar\gamma},_{\bar\lambda} + g_{\bar\gamma\bar\alpha,\bar\beta},_{\bar\lambda} - g_{\bar\beta\bar\gamma,\bar\alpha},_{\bar\lambda})/2,$$

$$R^{\bar\mu}_{\bar\beta\bar\gamma\bar\lambda} = g^{\bar\alpha\bar\mu}(g_{\bar\lambda\bar\alpha,\bar\beta},_{\bar\gamma} - g_{\bar\beta\bar\lambda,\bar\alpha},_{\bar\gamma} - g_{\bar\gamma\bar\alpha,\bar\beta},_{\bar\lambda} + g_{\bar\beta\bar\gamma,\bar\alpha},_{\bar\lambda})/2, \tag{4.7}$$

$$g_{\bar\nu\bar\mu}R^{\bar\mu}_{\bar\beta\bar\gamma\bar\lambda} = g_{\bar\nu\bar\mu}g^{\bar\alpha\bar\mu}(g_{\bar\lambda\bar\alpha,\bar\beta},_{\bar\gamma} - g_{\bar\beta\bar\lambda,\bar\alpha},_{\bar\gamma} - g_{\bar\gamma\bar\alpha,\bar\beta},_{\bar\lambda} + g_{\bar\beta\bar\gamma,\bar\alpha},_{\bar\lambda})/2, \tag{4.8}$$

$$R_{\bar\nu\bar\beta\bar\gamma\bar\lambda} = (g_{\bar\lambda\bar\nu,\bar\beta},_{\bar\gamma} - g_{\bar\beta\bar\lambda,\bar\nu},_{\bar\gamma} - g_{\bar\gamma\bar\nu,\bar\beta},_{\bar\lambda} + g_{\bar\beta\bar\gamma,\bar\nu},_{\bar\lambda})/2.$$

The above is the completely covariant curvature tensor. One then observes the following results follow trivially:

$$R_{\bar{\nu}\bar{\beta}\bar{\gamma}\bar{\lambda}} = -R_{\bar{\beta}\bar{\nu}\bar{\gamma}\bar{\lambda}} = -R_{\bar{\nu}\bar{\beta}\bar{\lambda}\bar{\gamma}} = R_{\bar{\gamma}\bar{\lambda}\bar{\nu}\bar{\beta}}, \tag{4.9}$$

$$0 = R_{\bar{\nu}\bar{\beta}\bar{\gamma}\bar{\lambda}} + R_{\bar{\nu}\bar{\lambda}\bar{\beta}\bar{\gamma}} + R_{\bar{\nu}\bar{\gamma}\bar{\lambda}\bar{\beta}}. \tag{4.10}$$

Equations (4.9) and (4.10) are tensor equations and so hold in all frames,

$$R_{\nu\beta\gamma\lambda} = -R_{\beta\nu\gamma\lambda} = -R_{\nu\beta\lambda\gamma} = R_{\gamma\lambda\nu\beta}, \tag{4.11}$$

$$0 = R_{\nu\beta\gamma\lambda} + R_{\nu\lambda\beta\gamma} + R_{\nu\gamma\lambda\beta}. \tag{4.12}$$

4.4   Ricci Tensor, Bianchi Identity, Einstein Tensor

The Ricci tensor is defined as follows:

$$R_{\beta\lambda} \equiv R^{\mu}{}_{\beta\mu\lambda}. \tag{4.13}$$

Note the contraction keeps the first and third covariant indexes. Other contractions yield zero or no additional information. It is easy to show, by starting in an inertial frame, that due to the symmetry properties of the curvature tensor,

$$R^{\mu}{}_{\beta\lambda\mu} = -R^{\mu}{}_{\beta\mu\lambda}, \quad 0 = R^{\mu}{}_{\mu\beta\lambda}. \tag{4.14}$$

As the metric tensor is symmetric, and the order of partial differentiation is unimportant, one can use Eqs. (4.7) and (4.11), rename summed over indexes, and show that this tensor is symmetric. Begin in a locally invariant frame,

$$R^{\bar{\mu}}{}_{\bar{\beta}\bar{\mu}\bar{\lambda}} = g^{\bar{\alpha}\bar{\mu}}(g_{\bar{\lambda}\bar{\alpha},\bar{\beta}}{}_{,\bar{\mu}} - (g_{\bar{\beta}\bar{\lambda},\bar{\alpha}}{}_{,\bar{\mu}} + g_{\bar{\mu}\bar{\alpha},\bar{\beta}}{}_{,\bar{\lambda}}) + g_{\bar{\beta}\bar{\mu},\bar{\alpha}}{}_{,\bar{\lambda}})/2,$$

$$R^{\bar{\mu}}{}_{\bar{\lambda}\bar{\mu}\bar{\beta}} = g^{\bar{\alpha}\bar{\mu}}(g_{\bar{\beta}\bar{\alpha},\bar{\lambda}}{}_{,\bar{\mu}} - (g_{\bar{\lambda}\bar{\beta},\bar{\alpha}}{}_{,\bar{\mu}} + g_{\bar{\mu}\bar{\alpha},\bar{\lambda}}{}_{,\bar{\beta}}) + g_{\bar{\lambda}\bar{\mu},\bar{\alpha}}{}_{,\bar{\beta}})/2,$$

$$R_{\bar{\beta}\bar{\lambda}} - R_{\bar{\lambda}\bar{\beta}} = g^{\bar{\alpha}\bar{\mu}}(g_{\bar{\lambda}\bar{\alpha},\bar{\beta}}{}_{,\bar{\mu}} + g_{\bar{\beta}\bar{\mu},\bar{\alpha}}{}_{,\bar{\lambda}} - g_{\bar{\beta}\bar{\alpha},\bar{\lambda}}{}_{,\bar{\mu}} - g_{\bar{\lambda}\bar{\mu},\bar{\alpha}}{}_{,\bar{\beta}})/2$$

$$= g^{\bar{\alpha}\bar{\mu}}(g_{\bar{\lambda}\bar{\mu},\bar{\beta}}{}_{,\bar{\alpha}} + g_{\bar{\beta}\bar{\alpha},\bar{\mu}}{}_{,\bar{\lambda}} - g_{\bar{\beta}\bar{\alpha},\bar{\lambda}}{}_{,\bar{\mu}} - g_{\bar{\lambda}\bar{\mu},\bar{\alpha}}{}_{,\bar{\beta}})/2 = 0.$$

The last equation is a tensor equation and holds in all frames:

$$R_{\beta\lambda} = R_{\lambda\beta}. \tag{4.15}$$

From the Ricci tensor an invariant, the Ricci scalar may be formed,

$$R = g^{\mu\nu}R_{\mu\nu} = g^{\mu\nu}R^{\xi}{}_{\mu\xi\nu} = g^{\mu\nu}g^{\chi\xi}R_{\chi\mu\xi\nu}. \tag{4.16}$$

The following contraction may also prove useful:

$$g^{\mu\nu} R^{\chi}{}_{\mu\beta\nu} = R^{\chi}{}_{\beta}. \tag{4.17}$$

For an inertial frame in free space, the Riemann curvature tensor is zero. This follows even if the metric tensor has non-constant elements because it is zero in rectangular coordinates. In cylindrical coordinates, the nonzero C symbols are $\Gamma^{\bar{2}'}_{\bar{1}'\bar{2}'} = 1/\rho$ and $\Gamma^{\bar{1}'}_{\bar{2}'\bar{2}'} = -\rho$. Notice that these C symbols have nonzero derivatives. Zero curvature will be explicitly demonstrated by calculating all possible nonzero elements of the Ricci tensor: $R_{\bar{1}'\bar{1}'}$, $R_{\bar{2}'\bar{2}'}$ and $R_{\bar{1}'\bar{2}'}$. In non-rectangular coordinates, one must use Eq. (4.5),

$$R_{\bar{\mu}'\bar{\mu}'} = R^{\bar{\xi}'}{}_{\bar{\mu}'\bar{\xi}'\bar{\mu}'} = \Gamma^{\bar{\chi}'}_{\bar{\mu}'\bar{\mu}'}\Gamma^{\bar{\xi}'}_{\bar{\xi}'\bar{\chi}'} - \Gamma^{\bar{\chi}'}_{\bar{\mu}'\bar{\xi}'}\Gamma^{\bar{\xi}'}_{\bar{\mu}'\bar{\chi}'} + \Gamma^{\bar{\xi}'}_{\bar{\mu}'\bar{\mu}',\bar{\xi}'} - \Gamma^{\bar{\xi}'}_{\bar{\mu}'\bar{\xi}',\bar{\mu}'},$$

$$R_{\bar{1}'\bar{1}'} = -\Gamma^{\bar{2}'}_{\bar{1}'\bar{2}'}\Gamma^{\bar{2}'}_{\bar{1}'\bar{2}'} - \Gamma^{\bar{2}'}_{\bar{1}'\bar{2}',\bar{1}'} = -\rho^{-2} - \rho^{-1}{}_{,\rho} = 0,$$

$$R_{\bar{2}'\bar{2}'} = \Gamma^{\bar{1}'}_{\bar{2}'\bar{2}'}\Gamma^{\bar{2}'}_{\bar{2}'\bar{1}'} - \Gamma^{\bar{1}'}_{\bar{2}'\bar{2}'}\Gamma^{\bar{2}'}_{\bar{2}'\bar{1}'} - \Gamma^{\bar{2}'}_{\bar{2}'\bar{1}'}\Gamma^{\bar{1}'}_{\bar{2}'\bar{2}'} + \Gamma^{\bar{1}'}_{\bar{2}'\bar{2}',\bar{1}'} = 1 - 1 = 0,$$

$$R_{\bar{1}'\bar{2}'} = \Gamma^{\bar{2}'}_{\bar{1}'\bar{2}'}\Gamma^{\bar{\xi}'}_{\bar{\xi}'\bar{2}'} - \Gamma^{\bar{2}'}_{\bar{2}'\bar{1}'}\Gamma^{\bar{2}'}_{\bar{2}'\bar{2}'} - +\Gamma^{\bar{2}'}_{\bar{2}'\bar{1}',\bar{2}'} - \Gamma^{\bar{\xi}'}_{\bar{\xi}'\bar{1}',\bar{2}'} = 0.$$

In Problem 3 it is shown that if constrained to a cylindrical surface where ρ = constant, the C symbols need recalculating, but there is still no curvature. In the case of spherical coordinates, the curvature vanishes in the unconstrained free space of an inertial frame, but not if constrained to a spherical surface.

Return to Eq. (4.8) and take a partial derivative,

$$R_{\bar{\nu}\bar{\beta}\bar{\gamma}\bar{\lambda}} = (g_{\bar{\lambda}\bar{\nu},\bar{\beta}\,,\bar{\gamma}} - g_{\bar{\beta}\bar{\lambda},\bar{\nu}\,,\bar{\gamma}} - g_{\bar{\gamma}\bar{\nu},\bar{\beta}\,,\bar{\lambda}} + g_{\bar{\beta}\bar{\gamma},\bar{\nu}\,,\bar{\lambda}})/2,$$

$$R_{\bar{\nu}\bar{\beta}\bar{\gamma}\bar{\lambda},\bar{\mu}} = (g_{\bar{\lambda}\bar{\nu},\bar{\beta}\,,\bar{\gamma}\,,\bar{\mu}} - g_{\bar{\beta}\bar{\lambda},\bar{\nu}\,,\bar{\gamma}\,,\bar{\mu}} - g_{\bar{\gamma}\bar{\nu},\bar{\beta}\,,\bar{\lambda}\,,\bar{\mu}} + g_{\bar{\beta}\bar{\gamma},\bar{\nu}\,,\bar{\lambda}\,,\bar{\mu}})/2,$$

$$R_{\bar{\nu}\bar{\beta}\bar{\mu}\bar{\gamma},\bar{\lambda}} = (g_{\bar{\gamma}\bar{\nu},\bar{\beta}\,,\bar{\mu}\,,\bar{\lambda}} - g_{\bar{\beta}\bar{\gamma},\bar{\nu}\,,\bar{\mu}\,,\bar{\lambda}} - g_{\bar{\mu}\bar{\nu},\bar{\beta}\,,\bar{\gamma}\,,\bar{\lambda}} + g_{\bar{\beta}\bar{\mu},\bar{\nu}\,,\bar{\gamma}\,,\bar{\lambda}})/2,$$

$$R_{\bar{\nu}\bar{\beta}\bar{\lambda}\bar{\mu},\bar{\gamma}} = (g_{\bar{\mu}\bar{\nu},\bar{\beta}\,,\bar{\lambda}\,,\bar{\gamma}} - g_{\bar{\beta}\bar{\mu},\bar{\nu}\,,\bar{\lambda}\,,\bar{\gamma}} - g_{\bar{\lambda}\bar{\nu},\bar{\beta}\,,\bar{\mu}\,,\bar{\gamma}} + g_{\bar{\beta}\bar{\lambda},\bar{\nu}\,,\bar{\mu}\,,\bar{\gamma}})/2,$$

$$0 = R_{\bar{\nu}\bar{\beta}\bar{\gamma}\bar{\lambda},\bar{\mu}} + R_{\bar{\nu}\bar{\beta}\bar{\mu}\bar{\gamma},\bar{\lambda}} + R_{\bar{\nu}\bar{\beta}\bar{\lambda}\bar{\mu},\bar{\gamma}} = R_{\bar{\nu}\bar{\beta}\bar{\gamma}\bar{\lambda};\bar{\mu}} + R_{\bar{\nu}\bar{\beta}\bar{\mu}\bar{\gamma};\bar{\lambda}} + R_{\bar{\nu}\bar{\beta}\bar{\lambda}\bar{\mu};\bar{\gamma}}.$$

The last step follows because in these coordinates, the partial derivative of a tensor is the covariant derivative. Thus, the above holds in any frame and is known as the Bianchi identity,

$$0 = R_{\nu\beta\gamma\lambda;\mu} + R_{\nu\beta\mu\gamma;\lambda} + R_{\nu\beta\lambda\mu;\gamma}. \tag{4.18}$$

Since the covariant derivative of the metric tensor is zero, we have from the Bianchi identity,

$$0 = g^{\nu\gamma}(R_{\nu\beta\gamma\lambda;\mu} + R_{\nu\beta\mu\gamma;\lambda} + R_{\nu\beta\lambda\mu;\gamma})$$
$$= (g^{\nu\gamma}R_{\nu\beta\gamma\lambda});_\mu + (g^{\nu\gamma}R_{\nu\beta\mu\gamma});_\lambda + (g^{\nu\gamma}R_{\nu\beta\lambda\mu});_\gamma.$$

Using Eqs. (4.11) and (4.13)–(4.18), this leads to

$$0 = R^\gamma{}_{\beta\gamma\lambda;\mu} + R^\gamma{}_{\beta\mu\gamma;\lambda} + R^\gamma{}_{\beta\lambda\mu;\gamma} = R_{\beta\lambda;\mu} - R^\gamma{}_{\beta\gamma\mu;\lambda} + R^\gamma{}_{\beta\lambda\mu;\gamma}$$
$$= R_{\beta\lambda;\mu} - R_{\beta\mu;\lambda} + R^\gamma{}_{\beta\lambda\mu;\gamma} = g^{\beta\lambda}(R_{\beta\lambda;\mu} - R_{\beta\mu;\lambda} - R^\gamma{}_{\beta\mu\lambda;\gamma})$$
$$= R_{;\mu} - R^\lambda{}_{\mu;\lambda} - R^\gamma{}_{\mu;\gamma} = \delta^\gamma{}_\mu R_{;\gamma} - 2R^\gamma{}_{\mu;\gamma}$$
$$= (\delta^\gamma{}_\mu R - 2R^\gamma{}_\mu);_\gamma = g^{\mu\nu}(\delta^\gamma{}_\mu R - 2R^\gamma{}_\mu);_\gamma = (g^{\gamma\nu}R - 2R^{\gamma\nu});_\gamma.$$

The Einstein tensor $G^{\gamma\nu}$ is defined in terms of the curvature:

$$G^{\gamma\nu} \equiv R^{\gamma\nu} - g^{\gamma\nu}R/2 = G^{\nu\gamma}, \tag{4.19}$$

$$G^{\gamma\nu};_\gamma = 0. \tag{4.20}$$

It is easy to show

$$G_{\gamma\nu} \equiv R_{\gamma\nu} - g_{\gamma\nu}R/2 = G_{\nu\gamma}, \tag{4.21}$$

$$G_{\gamma\nu};_\gamma = 0. \tag{4.22}$$

This tensor is symmetric, and is an essential quantity for solving problems involving gravity.

Problems

1. Show that $g^{\mu\nu}g_{\nu\chi,\beta} = -g^{\mu\nu}{}_{,\beta}\,g_{\nu\chi}$, $g^{\mu\nu}{}_{,\beta} = -(\Gamma^\mu{}_{\chi\beta}g^{\chi\nu} + \Gamma^\nu{}_{\chi\beta}g^{\chi\mu})$, and $\Gamma^\nu{}_{\mu\nu} = g^{\nu\chi}g_{\nu\chi,\mu}/2$.
2. Show that there are only 20 independent elements of $R_{\alpha\beta\mu\nu}$ that can be nonzero. Show that the contraction $R_{\mu\nu} = R^\chi{}_{\mu\chi\nu}$ is the only independent contraction of the Riemann curvature tensor.
3. Calculate the C symbols and curvature R in cylindrical coordinates in an inertial frame when $x^1 = \rho = a$ constant. Repeat the calculation for spherical coordinates. In that case, show that in unconstrained space

the curvature vanishes, but on the surface of a sphere, where $r = a$ constant, the curvature is finite.

4. In spherical coordinates, the Wormhole metric has nonzero elements: $g_{00} = -1$, $g_{rr} = 1$, $g_{\theta\theta} = a^2 + (r)^2$ and $g_{\phi\phi} = \sin^2\theta g_{\theta\theta}$, where a is a constant. Calculate the curvature R and show that it is negative.

5. Evaluate $G^{\mu\nu}{}_{,\mu}$ and $G_{\mu\nu,\mu}$ in terms of the metric, Ricci tensor $R_{\mu\nu}$, $R^{\mu\nu}$ and the C symbols.

6. It shall be seen that Einstein's equation for the metric is $G_{\mu\nu} + \Lambda g_{\mu\nu} = 8\pi T_{\mu\nu}$, where $G_{\mu\nu}$ is the Einstein tensor and $T_{\mu\nu}$ is the energy–momentum tensor. The factor 8π is required for agreement with Newtonian physics, where the latter is valid. The constant Λ is called the cosmological constant. It can be neglected for solar system problems. Einstein introduced it to make a static universe. When data showed an expanding universe, it was set to zero and Einstein called it his greatest error. As shall be seen, the universe is expanding too rapidly for $\Lambda = 0$. It must be positive and is now the dominant contribution to the energy in the universe. In vacuum $T_{\mu\nu} = 0$. What is R? In cosmology, a useful model is to describe the universe as a perfect fluid, for which $T_{\mu\nu} \neq 0$. On a solar system scale, earth can be thought to be moving in vacuum under the influence of the metric set up by the sun and $T_{\mu\nu} = 0$. Show that it is impossible for the worm hole metric of Problem 4 to exist in vacuum? Calculate all the elements of $T_{\mu\nu}$ for that metric including the cosmological constant.

7. The Robertson–Walker metric provides a good description of the universe,

$$(d\tau)^2 = (dt)^2 - Q^2(t)\left[\frac{(dr)^2}{1 - k(r)^2} + (r)^2([d\theta]^2 + \sin^2\theta[d\phi]^2)\right],$$

where $Q(t)$ is the universal scale factor, and the constant k is the curvature. They are obtained from experimental data. All observers use the same cosmic time t and any origin yields the same physics. Find the curvature R if we live in an expanding universe where $Q(t) > 0$ and $\frac{dQ(t)}{dt} > 0$.

8. For the metric of Problem 7, calculate the elements of the Einstein tensor $G_{\mu\nu}$.

9. For the metric of Problem 3.6, the nonzero C symbols were calculated. It will be seen that such a metric is that found by Schwarzschild for a planet moving in the sun's gravity. He derived the proper forms for

$\Phi(r)$ and $\Delta(r)$,

$$(d\tau)^2 = \exp[2\Phi(r)](dt)^2 - [\exp[2\Delta(r)](dr)^2 + (r)^2([d\theta]^2 + \sin^2\theta[d\phi]^2)].$$

Show that for this metric there are only six independent nonzero values of $R_{\mu\xi\chi\nu}$, and calculate them.

10. Calculate the elements of the Ricci tensor $R_{\mu\nu}$ from the results of Problem 9. Then calculate the curvature R and Einstein tensor $G_{\mu\nu}$.

Chapter 5

Gravity and General Relativity

5.1 Review of Newtonian Gravity

Gravitation, the weakest force, is due to, and acts on, all forms of energy. In Newtonian gravity, the relativistic mass M is the pertinent variable. The other known forces, explained by quantum mechanics, are due to the exchange of particles called force carriers. The weak force is experienced by neutrinos and all particles with rest mass. The force carriers are the $W^{\pm}$ and Z^0 bosons. Photons are the electromagnetic force carriers. Particles, with charge or higher order moments, experience this interaction. Particles made up of quarks, like pions, protons, and neutrons, experience the strong force, whose carriers are gluons. It should come as no surprise, that when gravity and quantum mechanics are connected, the graviton, see Chapter 7, becomes the force carrier. The theories of Newton and Einstein do not include such concepts.

Newton's theory of gravity is very similar to the classical theory of electrostatics. Students are usually more familiar with mathematics, like the Gauss law and the divergence theorem, from electrostatics. These two theories are compared, so that the desired gravitational result is simply obtained. For this short section, MKS units are used, so the equations will be familiar.

Both theories postulate an action at a distance force on each of two point particles $(1, 2)$, at positions $\vec{r}_{1,2}$. If $\vec{r}$ points from body 2 to body 1, with r being the distance between them, the forces on the bodies are as follows:

$$\vec{F}_{E,G}(1) = M(1)\vec{a}(1) = N_{E,G}f_{E,G}(1)f_{E,G}(2)\vec{r}/r^3 = -\vec{F}(2) = -M(2)\vec{a}(2),$$

$$N_E = (4\pi\epsilon_0)^{-1}, \quad f_E = Q, \quad N_G = -G, \quad f_G = M.$$

In electrostatics, charges Q can be positive or negative, and like charges repel, while unlike charges attract. Mass M is always positive, and the gravitational force is attractive. The normalization constants N_E and N_G reflect the strength of the forces.

In electrostatics, an electric field $\vec{E} \equiv -\vec{\nabla}\Psi_E$ can be defined as the negative gradient of the electrostatic potential Ψ_E. As the gravitational force has the same spatial form, the same may be done $\vec{G} \equiv -\vec{\nabla}\Psi_G$. Thus,

$$\vec{E}(1) \equiv \vec{F}_E(1)/Q(1) = (4\pi\epsilon_0)^{-1}Q(2)\vec{r}/r^3 = -\vec{\nabla}\Psi_E(1),$$

$$\Psi_E(1) = (4\pi\epsilon_0)^{-1}Q(2)/r,$$

$$\vec{G}(1) \equiv \vec{F}_G(1)/M(1) = -GM(2)\vec{r}/r^3 = -\vec{\nabla}\Psi_G(1),$$

$$\Psi_G(1) = -GM(2)/r.$$

$\Psi_{E,G}(1)$ is the electrostatic, gravitational potential at $\vec{r}_1$ because of the presence of the body at $\vec{r}_2$. If there are many bodies, $\Psi_{E,G}$ is the potential at a point in space, due to charges or masses at other points.

Due to the inverse square nature of the force, there is a Gauss law for both forces. Making use of the divergence theorem,

$$\int_V \vec{\nabla} \cdot \vec{\nabla}\Psi_{E,G}dV = \int_S \vec{\nabla}\Psi_{E,G} \cdot \hat{n}dS = -\int_S -\vec{\nabla}\Psi_{E,G} \cdot \hat{n}dS$$

$$= -4\pi N_{E,G}f_{E,G(\text{tot})} = -4\pi N_{E,G}\int_V \rho_{Q,M}dV,$$

$$\nabla^2\Psi_E = -\rho_Q/\epsilon_0,$$

$$\nabla^2\Psi_G = 4\pi G\rho_M$$

$$= 4\pi\rho_M \quad \text{(in natural units)}. \tag{5.1}$$

In natural units, the mass density ρ_M has units of m^{-2}. This result will be useful in obtaining the metric element g_{00} when gravity is weak. In the above equations, $\hat{n}$ is the outward unit 3-vector normal to surface S bounding volume V. The sources of the potentials ρ_Q and ρ_M are the charge and mass densities within V. The quantities $f_{E(\text{tot})}$ and $f_{G(\text{tot})}$ are the total charge and mass inside the volume. For the volume integral, $\nabla^2\Psi$

is evaluated at all points inside V, and for the surface integral, $\vec{\nabla}\Psi$ is evaluated at all points on the surface S. Note that, for point charges or masses $f_{E,G}(i)$, the charge or mass density can be expressed in terms of the Dirac delta function,

$$\rho_{E,M} = \sum_i f_{E,G}(i)\delta(\vec{r} - \vec{r}(i)).$$

Gravity is a very weak interaction. At the surface of the sun, $(M/R)_s = 1.484 \times 10^3/0.696 \times 10^9 = 2.13 \times 10^{-6}$. At the surface of earth, $(M/R)_e = 3 \times 10^{-4}(M/R)_s$. For a white dwarf, with the mass of the sun, but a radius 100 times smaller than that of the sun, $(M/R)_{wd} \approx 2 \times 10^{-4}$. For a neutron star, $M_{ns} = 1.4M_s$, $R_{ns} \approx 14$ km $= 20 \times 10^{-6}R_s$, $(M/R)_n = 0.15$. Here, GR is needed for accuracy. Aside from such compact objects and black holes, GR effects are minute.

5.2 Weak Gravity in GR

The language used in Newtonian mechanics is inappropriate. One should not say that mass M is moving in the gravitational potential produced by other masses, but rather, all the energy in the universe has caused there to be a non-flat metric in which M moves. For the earth, the sun is the main cause of the non-flatness and the metric is stationary $g_{\mu\nu,0} = 0$. The earth moves slowly compared with the speed of light, so one can take $U^i = 0$. The GR equations of motion yield

$$0 = \frac{dU^\alpha}{d\tau} + \Gamma^\alpha_{\mu\nu}U^\mu U^\nu = \frac{dU^\alpha}{d\tau} + \Gamma^\alpha_{00}U^0 U^0$$

$$= \frac{dU^\alpha}{d\tau} + g^{\alpha\chi}[g_{0\chi,0} + g_{0\chi,0} - g_{00,\chi}]\left(\frac{dx^0}{d\tau}\right)^2 \bigg/ 2$$

$$= \frac{dU^\alpha}{d\tau} - g^{\alpha\chi}g_{00,\chi}\left(\frac{dt}{d\tau}\right)^2 \bigg/ 2. \tag{5.2}$$

Gravity is a weak force, unless you are in the vicinity of a truly massive, compact object. So whatever $g_{\mu\nu}$ is, it is very close to $\eta_{\mu\nu}$. When the term *weak gravity* is specifically used, it means one is not seeking an exact solution, but rather an approximate one. Second-order deviations from $\eta_{\mu\nu}$ are neglected. As there is only a slight change from $\eta_{\mu\nu}$, rectangular coordinates are required, and the following definitions hold:

$$g_{\mu\nu} \equiv \eta_{\mu\nu} + h_{\mu\nu}, \qquad |h_{\mu\nu}| \ll 1, \tag{5.3}$$

$$g^{\mu\nu} \equiv \eta^{\mu\nu} + f^{\mu\nu}, \qquad |f^{\mu\nu}| \ll 1, \text{ and solving for } f^{\mu\nu}, \tag{5.4}$$

$$g^{\mu\alpha} g_{\alpha\nu} = \delta^{\mu}{}_{\nu} = (\eta^{\mu\alpha} + f^{\mu\alpha})(\eta_{\alpha\nu} + h_{\alpha\nu})$$

$$\approx \eta^{\mu\alpha}\eta_{\alpha\nu} + \eta^{\mu\alpha}h_{\alpha\nu} + \eta_{\alpha\nu}f^{\mu\alpha},$$

$$\delta^{\mu}{}_{\nu} = \delta^{\mu}{}_{\nu} + \eta^{\mu\alpha}h_{\alpha\nu} + \eta_{\alpha\nu}f^{\mu\alpha},$$

$$f^{\mu\chi} = \eta^{\chi\nu}\eta_{\alpha\nu}f^{\mu\alpha} = -\eta^{\chi\nu}\eta^{\mu\alpha}h_{\alpha\nu} = -h^{\mu\chi}. \tag{5.5}$$

For the motion of the earth about the sun, Newtonian mechanics provides an adequate description. The GR prediction must agree with Newton, to lowest order. In Eq. (5.2), each value of α has to be examined,

$$0 \approx \frac{dU^0}{d\tau} - \eta^{0\chi}h_{00,\chi}\left(\frac{dt}{d\tau}\right)^2 \Big/ 2$$

$$= \frac{dU^0}{d\tau} + h_{00,0}\left(\frac{dt}{d\tau}\right)^2 \Big/ 2 = \frac{dU^0}{d\tau},$$

$$U^0 = \frac{dx^0}{d\tau} = \frac{dt}{d\tau} = K = 1, \quad t = \tau,$$

$$0 = \frac{dU^i}{d\tau} - \eta^{i\chi}h_{00,\chi}/2$$

$$\approx \frac{d^2x^i}{dt^2} - h_{00,i}/2 = a^i - h_{00,i}/2.$$

One immediately sees that acceleration, due to gravity, is the source of a non-flat metric.

The constant choice $K = 1$ leads to agreement with Newtonian gravity. Let M be the mass of a moving object and M' be the mass that provides the metric. The last equation yields

$$M h_{00,i}/2 = M a^i = -M\Psi_{G,i},$$

$$h_{00}/2 = -\Psi_G + K' = -\Psi_G, \tag{5.6}$$

$$h_{00} = -2\Psi_G = 2M'/r.$$

Taking $K' = 0$ makes $-2\Psi_G = h_{00} = 0$ at $r = \infty$. This is the obvious choice to provide agreement with Newtonian gravity.

From the discussion at the end of Section 5.1, the extremely small values of M'/r near the surface of the earth or the sun make h_{00} very small. This was required. Also, if we consider Newtonian circular orbits of typical particles, the typical speed $\bar{v}$ is

$$M\bar{v}^2/r = MM'/(r)^2, \quad \bar{v}^2 = M'/r = h_{00}/2 \ll 1.$$

So a region of weak gravity is also a region of small speeds. It's worth noting that h_{00} is a correction of order of the typical small speed squared.

5.3 Gravitational Red Shift

Now that g_{00} has been obtained in the context of GR, the question of how gravity affects clocks can be reconsidered. Here, it is done in a completely GR context. A review of the material in Section 2.8 will prove worthwhile. The present discussion works, not only for the weak gravity metric under consideration, but for future exact metrics.

So suppose, as in Fig. 5.1, a helium source at r_2 emits photons. The photons are observed by experimenters at rest at r_1 and r_2, having traveled

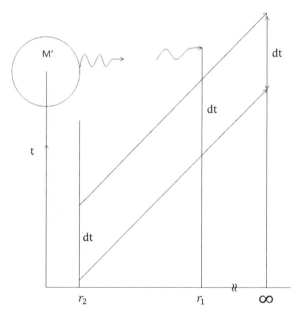

Fig. 5.1   Identical sources at $r_{1,2}$, in the metric set up by mass M', have their frequencies measured, by observers at rest at those positions. The frequency of the light that traveled from r_2 to r_1 is also determined.

in the metric set up by source mass M'. The observer at r_1 in weaker gravity also has a helium source. The desired prediction, that can be compared with measurement, is the ratio $\nu_{2,1}/\nu_{2,2} = \nu_{2,1}/\nu_{1,1}$. Here, the numerator is the frequency of the light from r_2 measured at r_1, and the denominator is the frequency from identical sources measured at the positions of the sources. The first method of predicting this ratio makes use of invariants,

$$[P^\mu]_{\text{photon}}[U_\mu]_{\text{obs.}} = [P^\mu]_{\text{photon}}\left[g_{\mu\nu}\frac{dx^\nu}{d\tau}\right]_{\text{obs.}}$$

$$= [P^0]_{\text{photon}}\left[g_{00}\frac{dt}{d\tau}\right]_{\text{obs.}} = [P^0]_{\text{photon}}([-g_{00}]_{\text{obs.}})^{1/2}$$

$$= h\nu_{2,2}(-g_{00}(2))^{1/2} = h\nu_{2,1}(-g_{00}(1))^{1/2}$$

$$= h\nu_{2,2}(1 - 2M'/r_2)^{1/2} = h\nu_{2,1}(1 - 2M'/r_1)^{1/2},$$

$$\nu_{2,1}/\nu_{1,1} = \nu_{2,1}/\nu_{2,2} = (1 - 2M'/r_2)^{1/2}(1 - 2M'/r_1)^{-1/2}$$

$$\approx 1 - M'(1/r_2 - 1/r_1). \tag{5.7}$$

A second way to get this result is to note, from Fig. 5.1, the world lines of successive wave crests. They travel from r_2 to r_1 and to a faraway point, along identical geodesics with speed c. The journey times are the same. The faraway at-rest observer is not experiencing gravity. That observer measures the proper-time period between crests $d\tau_\infty = dt$. This is the coordinate time period. Working backwards to $r_{1,2}$, along the world lines of the crests, one can see that dt is unchanged. However, the proper-time periods are the inverses of the frequencies,

$$\frac{d\tau_{2,2}}{d\tau_{2,1}} = \frac{dt_{2,2}}{dt_{2,1}}\left(\frac{-g_{00}(2)}{-g_{00}(1)}\right)^{1/2} = \left(\frac{-g_{00}(2)}{-g_{00}(1)}\right)^{1/2},$$

$$\frac{\nu_{2,1}}{\nu_{2,2}} = \frac{\nu_{2,1}}{\nu_{1,1}} = \left(\frac{-g_{00}(2)}{-g_{00}(1)}\right)^{1/2} \approx 1 - M'(1/r_2 - 1/r_1),$$

exactly as in Eq. (5.7).

Since $r_1 > r_2$, gravity is weaker at r_1. Then $\nu_{2,1}/\nu_{1,1} < 1$ and $\lambda_{2,1}/\lambda_{1,1} > 1$. This is known as the gravitational red shift. It is the same result obtained in Section 2.8 using energy conservation and relativistic mass. That worked for weak gravity. In addition, $d\tau_{2,2}/d\tau_{2,1} < 1$.

The proper-time tick rate for a clock depends on the position. It is greater for the clock experiencing weaker gravity.

5.4   Einstein's Field Equations

In Newton's theory, the gravitational potential Ψ_G is related to the source mass density ρ_M. This was seen in Eq. (5.1). For a source point mass M', the solution is $\Psi_G = -M'/r$, where r is the distance from the mass. Einstein's theory is based on curved spacetime, and one must determine the metric. The equation that relates the metric to the source must be covariant, a tensor equation. All observers would write the same equation, using their own coordinates. Instead of a mass density, that is related to the T^{00} element of the energy–momentum tensor, a covariant theory requires the entire tensor to be the source. So the correct equation would look like,

$$O(g^{\mu\nu}) = aT^{\mu\nu}, \tag{5.8}$$

where a is a constant, and O is a differential operator. The tensor properties of both sides of the above equation must be the same. Those are: each side is a symmetric tensor, and since momentum is conserved, each tensor has vanishing divergence $T^{\mu\nu}{}_{;\nu} = 0$. The tensor $T^{\mu\nu}$ is discussed in detail later, when it is needed. In the solar system, planets move in vacuum, in the metric set up by the sun. Thus $T^{\mu\nu} = 0$, and the constant a doesn't matter.

For curved spacetime, operator O will at least contain terms with the metric tensor and its first and second partial derivatives. That's just what is found in the Ricci tensor $R^{\mu\nu}$. A nonzero $R^{\mu}{}_{\nu\xi\chi}$ somewhere, means finite curvature, and the presence of gravity. Another way to say this is, at the position of the earth, it may be that $R^{\mu\nu} = 0$. However, at least one of the elements of the curvature tensor $R^{\mu}{}_{\nu\xi\chi} \neq 0$. Einstein struggled with this problem for many years before finally finding the simplest form,

$$O(g^{\mu\nu}) = R^{\mu\nu} + a'g^{\mu\nu}R + \Lambda g^{\mu\nu} = aT^{\mu\nu},$$

where a' and Λ are constants.

The constant Λ is known as the cosmological constant. It must be determined by observation. Einstein originally included it so that the universe would be static. When an expanding universe was discovered, he discarded it, considering its inclusion his greatest error. Present-day cosmology requires it. It is the driver of an accelerating expansion, that will never cease. As the universe gets bigger, its effect gets stronger. However, for solar system predictions, it is extremely small, and may be neglected.

From momentum conservation,

$$0 = T^{\mu\nu}{}_{;\nu} = [(R^{\mu\nu} + a'g^{\mu\nu}R + \Lambda g^{\mu\nu})_{;\nu}] = [(R^{\mu\nu} + a'g^{\mu\nu}R)_{;\nu}].$$

However, when the Einstein tensor $G^{\mu\nu}$ was derived, Eqs. (4.19) and (4.20) yielded,

$$G^{\mu\nu} = R^{\mu\nu} - g^{\mu\nu}R/2, \quad G^{\mu\nu}{}_{;\nu} = 0.$$

So take $a' = -1/2$, then $O(g^{\mu\nu}) = G^{\mu\nu} + \Lambda g^{\mu\nu}$. The Einstein field equations become

$$G^{\mu\nu} + \Lambda g^{\mu\nu} = aT^{\mu\nu} = 8\pi T^{\mu\nu}. \tag{5.9}$$

The value $a = 8\pi$ is required, so that the predictions of GR agree with the Newtonian predictions, where the latter are applicable. The reader is led through this calculation in Problem 3. In the empty space of our solar system, there is no energy–momentum tensor,

$$0 = T^{\mu\nu} = G^{\mu\nu} = R^{\mu\nu}. \tag{5.10}$$

More complicated field equations have been proposed, but Eq. (5.9), selected by Einstein for its simplicity and beauty, has withstood every experimental test.

5.5   Schwarzschild Solution

Almost immediately after Einstein introduced the field equations, K. Schwarzschild found an exact solution. An English translation of his paper is available, see Schwarzschild (1916). The case in point was the one that was considered with weak gravity, the metric produced by a static, spherically symmetric, massive object in vacuum. The metric tensor won't depend on t, but will depend on $\vec{r}$ and $d\vec{r}$, such that it has rotational invariance. Time independence leads to energy conservation, and rotational invariance leads to conservation of certain angular momentum components. Thus, it pays to work in spherical coordinates. So for the rest of this chapter, r^p means $(r)^p$, and not the pth component of the position vector.

Try the most general rotationally invariant form for the proper-time element,

$$(d\tau)^2 = -g_{\mu\nu}dx^\mu dx^\nu$$

$$\equiv A(r)(dt)^2 - 2B(r)(\vec{r}\cdot d\vec{r})dt - C(r)(\vec{r}\cdot d\vec{r})^2 - D(r)d\vec{r}\cdot d\vec{r}$$

$$= A(r)(dt)^2 - 2B(r)rdrdt - [C(r)r^2 + D(r)](dr)^2$$

$$- D(r)r^2[(d\theta)^2 + (\sin\theta d\phi)^2]. \tag{5.11}$$

One can eliminate the $dtdr$ term with the following transformation:

$$t \equiv t' - E(r), \qquad dE(r) \equiv -rdrB(r)/A(r). \tag{5.12}$$

It is then easy to show, see Problem 4, that this leads to

$$(d\tau)^2 = A(r)(dt')^2 - [(rB(r))^2/A(r) + C(r)r^2 + D(r)](dr)^2$$

$$- D(r)r^2[(d\theta)^2 + \sin^2\theta(d\phi)^2]$$

$$\equiv A(r)(dt')^2 - F(r)(dr)^2 - r^2D(r)[(d\theta)^2 + (\sin\theta d\phi)^2]. \tag{5.13}$$

A final transform redefines r, and allows the proper time to be cast in a form where the metric tensor is diagonal,

$$r'^2 \equiv r^2 D(r), \tag{5.14}$$

$$(d\tau)^2 \equiv \exp[2\Phi(r')](dt')^2 - \exp[2\Delta(r')](dr')^2$$

$$- r'^2[(d\theta)^2 + (\sin\theta d\phi)^2]. \tag{5.15}$$

From now on the primes will be dropped. The last form has been encountered in many of the problems. Since the metric tensor is diagonal, $g^{\mu\mu} = 1/g_{\mu\mu}$. In Problem 3.6, the C symbols were calculated, and the nonzero ones are as follows:

$$\Gamma^r_{rr} = \Delta_{,r},$$

$$\sin^2\theta\Gamma^r_{\theta\theta} = -r\exp[-2\Delta]\sin^2\theta = \Gamma^r_{\phi\phi},$$

$$\Gamma^r_{tt} = \exp[-2\Delta]\exp[2\Phi]\Phi_{,r},$$

$$\Gamma^\theta_{r\theta} = \Gamma^\theta_{\theta r} = r^{-1}, \qquad \Gamma^\theta_{\phi\phi} = -\sin\theta\cos\theta,$$

$$\Gamma^\phi_{r\phi} = \Gamma^\phi_{\phi r} = r^{-1}, \qquad \Gamma^\phi_{\phi\theta} = \Gamma^\phi_{\theta\phi} = \cot\theta,$$

$$\Gamma^t_{rt} = \Gamma^t_{tr} = \Phi_{,r}.$$

$$\tag{5.16}$$

In Problem 4.10, the nonzero Einstein tensor elements were found:

$$G_{tt} = r^{-2} \exp[2\Phi][r(1 - \exp[-2\Delta])]_{,r},$$

$$G_{\theta\theta} = (r)^2 \exp[-2\Delta][\Phi_{,r}\,_{,r} + (\Phi_{,r})^2 - \Phi_{,r}\Delta_{,r} + (r)^{-1}(\Phi_{,r} - \Delta_{,r})],$$
$$(5.17)$$

$$G_{\phi\phi} = \sin^2\theta G_{\theta\theta},$$

$$G_{rr} = -r^{-2}(\exp[2\Delta] - 1) + 2r^{-1}\Phi_{,r}.$$

In empty space $G_{\mu\nu} = 0$ because $T_{\mu\nu} = 0$,

$$0 = G_{tt} = r^{-2} \exp[2\Phi][r(1 - \exp[-2\Delta])]_{,r}$$

$$0 = [r(1 - \exp[-2\Delta])]_{,r},$$

$$b = r(1 - \exp[-2\Delta]),$$

$$\exp[-2\Delta] = 1 - b/r,$$

$$0 = G_{rr} = -r^{-2}(\exp[2\Delta] - 1) + 2(r)^{-1}\Phi_{,r}, \text{ so,}$$

$$2r^{-1}\Phi_{,r} = r^{-2}(1/(1 - b/r) - 1) = r^{-2}(b/r)/(1 - b/r),$$

$$2\Phi_{,r} = r^{-2}b/(1 - b/r) = (\ln[1 - b/r])_{,r},$$

$$2\Phi = \ln[(1 - b/r)b'], \qquad \exp[2\Phi] = b'(1 - b/r).$$

The last equation means $g_{00} = -b'(1 - b/r) = b'(-1 + b/r)$. However, the weak gravity case yielded $g_{00} = -1 + 2M'/r$. The Schwarzschild and weak gravity results must agree in the case that gravity is weak, so the constants are $b' = 1$ and $b = 2M'$. Here, M' is the relativistic mass of the metric source, e.g., the sun. Then,

$$\exp[2\Phi] = \exp[-2\Delta] = (1 - 2M'/r). \tag{5.18}$$

If $G_{\mu\nu} = 0$, then $R_{\mu\nu} = 0$. So at first it appears that the planets are not moving in the vacuum of curved space. However, not all the elements of $R_{\mu\nu\xi\beta} = g_{\mu\alpha}R^{\alpha}{}_{\nu\xi\beta} = 0$. There is curvature. The solution of Problem 4.9 indicates that there are six possible nonzero elements of the completely covariant curvature tensor. In particular, one can show $R_{0330} = 2M'/r^3$.

If the cosmological constant Λ was included, see Problem 5, it is still simple to calculate the functions $\exp[2\Phi]$ and $\exp[2\Delta]$. Then one can show that in the weak gravity approximation $M'/r \ll 1$, $\Lambda \ll 1$, there is a repulsive Newtonian force due to Λ. This makes sense, as the cosmological constant is driving the universal expansion. However, as will be seen in the problem, its contribution on a solar system scale cannot be observed.

5.6   Conserved Quantities: Massive Particles

Knowledge of the metric tells us what quantities, if any, are conserved. Such knowledge is very helpful in solving the equations of motion. Light has one constant of the motion, its speed. Equation (4.3) was obtained from parallel transport and for a particle of rest mass m, $d\tau \neq 0$ so it can be used instead of dq. Using $g_{\mu\alpha;\nu} = 0$, and renaming summed over indexes,

$$0 = W^\nu W^\mu{}_{;\nu} = U^\nu U^\mu{}_{;\nu} = P^\nu P^\mu{}_{;\nu}$$

$$= g_{\mu\alpha} P^\nu P^\mu{}_{;\nu} = P^\nu (g_{\mu\alpha} P^\mu){}_{;\nu}$$

$$= P^\nu P_\alpha{}_{;\nu} = P^\nu (P_{\alpha,\nu} - P_\beta \Gamma^\beta_{\alpha\nu})$$

$$= m\frac{dx^\nu}{d\tau} P_{\alpha,\nu} - P^\nu P_\beta \Gamma^\beta_{\alpha\nu} = m\frac{dP_\alpha}{d\tau} - P^\nu P_\beta \Gamma^\beta_{\alpha\nu}.$$

Then,

$$\frac{dP_\alpha}{d\tau} = P^\nu P_\beta \Gamma^\beta_{\alpha\nu}/m = P^\nu P_\beta g^{\beta\chi}(g_{\alpha\chi,\nu} + g_{\chi\nu,\alpha} - g_{\alpha\nu,\chi})/(2m)$$

$$= (P^\nu P^\chi g_{\alpha\chi,\nu} + P^\chi P^\nu [g_{\chi\nu,\alpha} - g_{\alpha\nu,\chi}])/(2m)$$

$$= P^\chi P^\nu (g_{\alpha\nu,\chi} + g_{\chi\nu,\alpha} - g_{\alpha\nu,\chi})/(2m)$$

$$= P^\chi P^\nu g_{\chi\nu,\alpha}/(2m). \qquad (5.19)$$

So if $g_{\chi\nu,\alpha} = 0$, then P_α is constant along the geodesic. In a stationary metric $g_{\chi\nu,0} = 0$ and P_0 is constant. In the case of weak gravity and low speeds, the total energy is constant.

The constancy of energy can be illustrated to lowest order in small quantities. Use the fact that $|\vec{P}| \ll m$ so that $h_{ij} P^i P^j / m^2$ and $h_{ii} |\vec{P}|^2 / m^2$ can be neglected,

$$m^2 = -g_{\mu\nu} P^\mu P^\nu$$

$$= -(-1 + h_{00})(P^0)^2 - [(1 + h_{ii})|\vec{P}|^2 + 2h_{ij} P^i P^j],$$

$$1 \approx (1 - h_{00})(P^0/m)^2 - (|\vec{P}|/m)^2$$

$$= (1 + 2M'/r)(P^0/m)^2 - (|\vec{P}|/m)^2,$$

$$P^0/m = (1 + (|\vec{P}|/m)^2)^{1/2}(1 + 2M'/r)^{-1/2}$$

$$\approx (1 + (|\vec{P}|/m)^2/2)(1 - M'/r),$$

$$P^0 \approx m - mM'/r + |\vec{P}|^2/(2m) = RE + PE + KE = E,$$

$$P_0 = g_{00} P^0 \approx -P^0 = -E, \text{ constant.}$$

Outside a spherically symmetric massive body there will be axial symmetry. One can find coordinates such that the $g_{\mu\nu}$ are independent of the angle about that axis. One can take the axis such that ϕ is that angle. Using the Schwarzschild metric for a slowly moving particle,

$$P_\phi = g_{\phi\phi}P^\phi$$

$$= (r\sin\theta)^2 m \frac{d\phi}{dt}$$

is constant. You recognize this relation as the conservation of an angular momentum component.

5.7 The Twin Problem From the Rocket Observer's View

There is a wonderful example that makes use of all the concepts thus far presented. It concerns the twin problem. That problem was solved, in Chapter 2, by the inertial twin I. Here, the way the accelerated twin R solves the problem with GR is presented. Of course, the final comparison of the clocks must agree with what was previously found.

Neglecting gravity, consider two identical twins I and R carrying identically constructed clocks. The clocks are synchronized at $t_I = t_R = 0$, when both are at $z_I = z_R = 0$. Observer I never experiences acceleration. She is an inertial observer, who can use SR, to describe the motion of R. She notes that R accelerates away, reverses acceleration, and slows to zero speed. R continues with the reverse acceleration, and speeds towards I. A final acceleration reversal brings R to rest at I's position. They compare clock times and R is younger.

R says she always experiences acceleration. During the four legs of the trip, each of duration T on the R clock, the acceleration is

$$\vec{a}_R = \begin{cases} g\hat{e}_z, & 0 \le t_R \le T, \text{ Leg 1}, \\ -g\hat{e}_z, & T \le t_R \le 2T, \text{ Leg 2}, \\ -g\hat{e}_z, & 2T \le t_R \le 3T, \text{ Leg 3}, \\ g\hat{e}_z, & 3T \le t_R \le 4T, \text{ Leg 4}. \end{cases} \tag{5.20}$$

R sees I depart, speeding up in the $-\hat{e}_z$-direction, then slowing to rest after the acceleration reverses. I speeds towards her with the reverse acceleration. After the last acceleration reversal, I is brought to rest at R's position.

Both are aware of the information in Eq. (5.20). This allows I to calculate, using SR, the trip time on her clock, as in Section 2.5. According to I, at the end of the trip, the ratio of the times on the clocks is

$$gt_{I,\text{tot}}/(gt_{R,\text{tot}}) = gt_{I,\text{tot}}/(4gT) = 4\sinh[gT]/(4gT). \qquad (5.21)$$

I finds that each leg of the trip has the same duration, so that $gt_I[T] = \sinh[gT]$. At the end of the trip, the twins are at rest at the same position, and can physically compare clocks. They verify Eq. (5.21). Also, I predicts that R's maximum speed is

$$V = \tanh[gT]. \qquad (5.22)$$

This occurs at the end of legs $1,3$. The maximum separation distance $2D$ was calculated in Problem 2.8,

$$
\begin{aligned}
gz_I[t_I] &= [(1 + [gt_I]^2)^{1/2} - 1] = [(1 - v^2)^{-1/2} - 1], \\
gz_I[t_R] &= [(1 + \sinh^2[gt_R])^{1/2} - 1] = \cosh[gt_R] - 1, \\
2gD &= 2gz_I[T] \\
&= 2[(1 - V^2)^{-1/2} - 1] = 2(\cosh[gT] - 1).
\end{aligned}
\qquad (5.23)
$$

In the above equations, z_I is the position of R, as determined by I, and t_I, t_R are the times on the clocks of I, R, when R is at that position.

Here, the journey of I is calculated by R. The latter, though experiencing acceleration, feels fixed at the origin. Dropping subscripts, R uses coordinate z for I's position when the time on the R clock is t. No use is made of any part of the calculations of I. In order to account for the acceleration, R finds herself in a complicated metric, and uses the equations of GR. As shall be seen, the durations on I's clock, as calculated by R, for legs 1 and 4, are different from legs 2 and 3. However, when the clocks are finally back in coincidence, the duration on each observer's clock is the same as that calculated by I.

The method of R. Perrin (1979) is followed with some changes and additions. Perrin noted that there aren't any treatments in the literature that use the full power of GR. Möller (1952) transforms the flat metric into a form that is valid for accelerated frames, but it leads to complicated geodesics.

Perrin proceeds as one does for the Schwarzschild metric. Assume the nonzero metric elements are $g_{00} = -\exp[\alpha]$ and $g_{33} = \exp[\beta]$, where α and

β are functions of z,

$$(d\tau)^2 = \exp[\alpha](dt)^2 - \exp[\beta](dz)^2. \tag{5.24}$$

R will calculate the duration $d\tau$ on the clock attached to I.

The nonzero C symbols are as follows:

$$\Gamma^0_{03} = \frac{\partial\alpha}{\partial z}/2 \equiv \alpha_{,3}/2 = \alpha_{,z}/2,$$

$$\Gamma^3_{33} = \beta_{,z}/2, \quad \Gamma^3_{00} = \exp[\alpha - \beta]\alpha_{,z}/2. \tag{5.25}$$

The Ricci Tensor $R_{\mu\nu}$ has the following nonzero elements,

$$R_{33} = -[\alpha_{,z}\,(\alpha_{,z} - \beta_{,z}\,)/2 + \alpha_{,z\,,z}\,]/2,$$

$$R_{00} = -R_{33}\exp[\alpha - \beta], \tag{5.26}$$

$$R = g^{00}R_{00} + g^{33}R_{33} = 2R_{33}\exp[-\beta].$$

The function β can be obtained in terms of α. Use Eq. (5.26) and that the motion takes place in free space,

$$R_{33} = 0, \quad \text{so},$$

$$\beta_{,z} = 2\alpha_{,z\,,z}/\alpha_{,z} + \alpha_{,z}, \tag{5.27}$$

$$\beta = 2\ln[|\alpha_{,z}|] + \alpha - \ln[K^2].$$

When gravity is weak, GR and Newtonian gravity agree. This occurs when $\alpha \ll 1$ and $\exp[\alpha] \approx 1 + \alpha$. R says I is moving, and has acceleration $a = \mp g$ when close to $z = 0$. R says her own acceleration is $\pm g$ at $z = 0$. Using the group of Eq. (5.6), R deduces α from I's motion,

$$-1 + h_{00} \approx g_{00} = -\exp[\alpha] \approx -(1 + \alpha), \quad h_{00} = -\alpha,$$

$$a = \mp g = h_{00,z}/2 = -\alpha_{,z}/2, \tag{5.28}$$

$$\alpha_{,z} = \pm 2g, \quad \alpha = \pm 2gz.$$

Then Eq. (5.27) yields, on choosing $K^2 = 4g^2$,

$$\beta = 2\ln[2g] + \alpha - \ln[4g^2] = \alpha, \quad \text{so},$$

$$g_{00} = -\exp[\pm 2gz] = -g_{33}, \quad \Gamma^0_{03} = \Gamma^3_{00} = \Gamma^3_{33} = \pm g. \tag{5.29}$$

The choice of the constant is the natural one. A complicated metric, satisfying all the physics is obtained, without dependence on an arbitrary constant.

An interesting observation can be made. Since the derivatives of the C symbols vanish, the curvature tensor is

$$R^{\mu}{}_{\beta\gamma\lambda} = \Gamma^{\nu}{}_{\lambda\beta}\Gamma^{\mu}{}_{\nu\gamma} - \Gamma^{\nu}{}_{\gamma\beta}\Gamma^{\mu}{}_{\nu\lambda}.$$

The C symbols vanish unless $\mu = 0, 3$ and all finite C symbols are equal. This yields that every element $R^{\mu}{}_{\beta\gamma\lambda} = 0$. Unlike the Schwarzschild problem, where $R_{\mu\nu} = 0$, but some elements $R^{\mu}{}_{\beta\gamma\lambda} \neq 0$, here there is no curvature. So observer R could attack this problem using the tools of SR.

There is no finite energy–momentum tensor, at some position, that is the source of the metric. Here, the complicated metric is necessary, for the observer experiencing acceleration. This is similar to, fictitious forces for accelerating observers, appearing in Newtonian mechanics. In the early days of relativity, the source of such metrics were a problem. Möller called them due to a non-permanent gravitational interaction. He noted that Einstein considered the source was acceleration of the distant "fixed" stars. However, there are no fixed stars. In fact, the farther away the stars are, the faster they are accelerating relative to any observer.

The metric gives rise to the following equations of motion:

$$0 = \frac{d^2 x^{\mu}}{d\tau^2} + \Gamma^{\mu}{}_{\nu\xi}\frac{dx^{\nu}}{d\tau}\frac{dx^{\xi}}{d\tau}$$

$$= \frac{d^2 t}{d\tau^2} \pm 2g\frac{dt}{d\tau}\frac{dz}{d\tau} \tag{5.30}$$

$$= \frac{d^2 z}{d\tau^2} \pm g\left[\left(\frac{dt}{d\tau}\right)^2 + \left(\frac{dz}{d\tau}\right)^2\right]. \tag{5.31}$$

Combining Eqs. (5.30) and (5.31) yields

$$\frac{d^2(z \pm t)}{d\tau^2} \pm g\left[\frac{d(z \pm t)}{d\tau}\right]^2 = 0. \tag{5.32}$$

These equations are easily solved,

$$f \equiv \frac{d(z \pm t)_{\pm}}{d\tau},$$

$$\frac{df}{d\tau} = \mp g f^2,$$

$$f^{-2}df = \mp g d\tau,$$

$$-f^{-1} = \mp g\tau - K^{\pm}_{\pm},$$

$$f = \frac{d(z \pm t)_\pm}{d\tau} = (K_\pm^\pm \pm g\tau)^{-1},$$

$$(z \pm t)_\pm = \pm \ln[|K_\pm^\pm \pm g\tau|]/g + C_\pm^\pm.$$

$$(5.33)$$

On the constants, the superscript $\pm$ refers to $z \pm t$, and the subscript $\pm$ refers to $\pm g$. Thus, solutions for z and t become

$$z_\pm = [(z + t)_\pm + (z - t)_\pm]/2$$

$$= A_\pm \pm \ln[|(K_\pm^+ \pm g\tau)(K_\pm^- \pm g\tau)|]/(2g), \qquad (5.34)$$

$$t_\pm = [(z + t)_\pm - (z - t)_\pm]/2$$

$$= B_\pm \pm \ln[|(K_\pm^+ \pm g\tau)/(K_\pm^- \pm g\tau)|]/(2g). \qquad (5.35)$$

The constants $A_\pm, B_\pm, K_\pm^\pm$ are determined from the boundary conditions.

Due to the absolute value of the log argument, one of the following possibilities holds:

$$\ln[|(K_\pm^+ \pm g\tau)(K_\pm^- \pm g\tau)|] = \ln[\pm(K_\pm^+ \pm g\tau)(K_\pm^- \pm g\tau)],$$

$$\ln[|(K_\pm^+ \pm g\tau)/(K_\pm^- \pm g\tau)|] = \ln[\pm(K_\pm^+ \pm g\tau)/(K_\pm^- \pm g\tau)].$$

$$(5.36)$$

The correct choice comes from the metric,

$$1 = \exp[\pm 2gz]\left[\left(\frac{dt}{d\tau}\right)^2 - \left(\frac{dz}{d\tau}\right)^2\right].$$

Thus,

$$\frac{dz}{d\tau} = \frac{K_\pm^- + K_\pm^+ \pm 2g\tau}{2(K_\pm^+ \pm g\tau)(K_\pm^- \pm g\tau)},$$

$$\frac{dt}{d\tau} = \frac{K_\pm^- - K_\pm^+}{2(K_\pm^+ \pm g\tau)(K_\pm^- \pm g\tau)},$$

$$\dot{z} \equiv \frac{dz}{dt} = \frac{dz}{d\tau} \Big/ \frac{dt}{d\tau} = \frac{K_\pm^- + K_\pm^+ \pm 2g\tau}{K_\pm^- - K_\pm^+}, \qquad (5.37)$$

$$1 = \frac{\exp[\pm 2gA_\pm]}{4} \frac{(K_\pm^- - K_\pm^+)^2 - (K_\pm^- + K_\pm^+ \pm 2g\tau)^2}{\pm(K_\pm^+ \pm g\tau)(K_\pm^- \pm g\tau)}$$

$$= \pm \exp[\pm 2gA_\pm](-1).$$

So $A_\pm = 0$, and the negative sign is required. Equations (5.34) and (5.35) now become

$$z_\pm = \pm \ln[-(K_\pm^+ \pm g\tau)(K_\pm^- \pm g\tau)]/(2g), \qquad (5.38)$$

$$t_\pm = B_\pm \pm \ln[-(K_\pm^+ \pm g\tau)/(K_\pm^- \pm g\tau)]/(2g). \qquad (5.39)$$

Initially, take $g_{33} = -g_{00} = \exp[2gz]$. The other initial boundary conditions for I's motion are: $\tau = t = z = \dot{z} = 0$. Then Eq. (5.37) yields $K_+^- = -K_+^+$, Eq. (5.39) yields $B_+ = 0$, and Eq. (5.38) yields $K_+^+ = 1$. At $t = T$, one gets

$$T = \ln[(1 + g\tau)/(1 - g\tau)]/(2g),$$

$$\exp[2gT] = (1 + g\tau)/(1 - g\tau),$$

$$g\tau[T] = (\exp[2gT] - 1)/(\exp[2gT] + 1) = \tanh gT = V, \qquad (5.40)$$

$$\dot{z}[T] = 2g\tau/(-2) = -V, \qquad (5.41)$$

$$z[T, g] = \ln[-(1 + g\tau)(-1 + g\tau)]/(2g)$$

$$= \ln[1 - V^2]/(2g) = \ln[(1 - V^2)^{1/2}]/g. \qquad (5.42)$$

Equations (5.40) and (5.41) agree with Eq. (5.22), the speed calculated by I. Equation (5.40) shows R is aging faster than I. However, they cannot compare clocks, as they are at different positions. They must wait until they are coincident, to compare times. The positive proper distance between R and I is calculated using Eq. (5.42),

$$L_P[T] = -\int_0^{z[T,g]} \exp[gz']dz' = (1 - \exp[gz[T, g]])/g$$

$$= [1 - (1 - V^2)^{1/2}]/g,$$

$$gL_P[T] = (\cosh[gT] - 1)/\cosh[gT] = gD/(1 + gD), \qquad (5.43)$$

$$2gD = 2(\cosh[gT] - 1).$$

This agrees with Eq. (5.23), I's prediction.

The rocket acceleration discontinuously changes to $-g$, and thus the metric changes discontinuously. Coordinates have no physical meaning. They can be translated or rotated to quite different values. The proper distance from R to I is a physical quantity. It cannot be translated or rotated to some new value, and won't change because the acceleration changed. Thus, the z coordinate will have to change discontinuously, from $z[T, g]$, Eq. 5.42, to $z[T, -g]$, such that the proper distance is unchanged. The starting spatial coordinate for the trip's second leg is obtained from

$$[1 - (1 - V^2)^{1/2}]/g = \int_{z[T,-g]}^{0} \exp[-gz]dz$$

$$= -(1/g)\exp[-gz]|_{z[T,-g]}^{0}$$

$$= (\exp[-gz[T,-g]] - 1)/g,$$

$$z[T,-g] = -\ln[2 - (1 - V^2)^{1/2}]/g$$

$$= -\ln[(2 - (1 - V^2)^{1/2})^2]/(2g). \qquad (5.44)$$

For the second leg, the starting values are $t = T$, $g\tau[T] = V$. The general solutions, Eqs. (5.37), (5.38), and (5.39) become

$$\dot{z} = \frac{K_-^- + K_-^+ - 2(g\tau - V)}{K_-^- - K_-^+},$$

$$z = -\ln[-(K_-^+ - (g\tau - V))(K_-^- - (g\tau - V))]/(2g), \qquad (5.45)$$

$$t - T = B_- - \ln[-(K_-^+ - (g\tau - V))/(K_-^- - (g\tau - V))]/(2g).$$

The constants are determined from the values of these variables at the start of the second leg,

$$\dot{z} = -V = (K_-^- + K_-^+)/(K_-^- - K_-^+),$$
$$K_-^- = -K_-^+(1 - V)/(1 + V); \qquad (5.46)$$

$$z = -\ln[(2 - (1 - V^2)^{1/2})^2]/(2g) = -\ln[-K_-^+K_-^-]/2g,$$

$$= -\ln[(K_-^+)^2(1 - V)/(1 + V)]/(2g), \qquad (5.47)$$

$$K_-^+ = (2 - (1 - V^2)^{1/2})[(1 + V)/(1 - V)]^{1/2};$$

$$0 = t - T = B_- - \ln[-K_-^+/K_-^-]/2g$$

$$= B_- - \ln[(1 + V)/(1 - V)]/2g, \qquad (5.48)$$

$$B_- = \ln[(1 + V)/(1 - V)]/2g.$$

At the end of leg 2, $t = 2T$. Equation (5.45), use of the above constants, and $\exp[2gT] = (1 + V)/(1 - V)$ yield

$$T = B_- - \ln[-(K_-^+ - (g\tau - V))/(K_-^- - (g\tau - V))]/(2g),$$

$$\frac{1 + V}{1 - V} = \frac{1 + V}{1 - V} \frac{(\frac{1-V}{1+V})^{1/2}(2 - (1 - V^2)^{1/2}) + (g\tau - V)}{(\frac{1+V}{1-V})^{1/2}(2 - (1 - V^2)^{1/2}) - (g\tau - V)},$$

$$g\tau = 2V(1 - V^2)^{-1/2} = 2\sinh[gT].$$

In leg 2, $g\tau$ advanced by $2\sinh[gT] - \tanh[gT]$. The total duration at this point $2\sinh[gT]$ is the expected value. Legs $3, 4$ just duplicate the times of legs $2, 1$. When the observers are coincident, the ratio of the total times on the clocks is

$$g\tau_{\text{tot}}/(gt_{\text{tot}}) = 4\sinh[gT]/(4gT),$$

This agrees with Eq. (5.21).

The velocity at the end of leg 2 is

$$K_-^- + K_-^+ = -K_-^+((1 - V)/(1 + V) - 1)$$

$$= 2VK_-^+/(1 + V)$$

$$= 2V(2(1 - V^2)^{-1/2} - 1) = 2(g\tau - V),$$

$$\dot{z}[2T] = \frac{K_-^- + K_-^+ - 2(g\tau - V)}{K_-^- - K_-^+} = 0.$$

This is the expected result. If I says R is at rest, then R must say I is at rest. The position at the end of leg 2 is

$$z[2T] = -\ln[-(K_-^+ - (g\tau - V))(K_-^- - (g\tau - V))]/(2g) \quad (5.49)$$

$$= -(1/g)\ln[2(1 - V^2)^{-1/2} - 1]. \quad (5.50)$$

This enables calculation of the proper distance

$$L_P[2T] = \int_{z[2T]}^0 \exp[-gz]dz = (\exp[-gz[2T]] - 1)/g$$

$$= 2[(1 - V^2)^{-1/2} - 1]/g = 2(\cosh[gT] - 1) = 2D.$$

This result is in agreement with Eq. (5.23), the distance calculated by I. It had to be since both observers are at rest at this point. Thus, they are in the same Lorentz frame.

This calculation differs from Perrin's calculation. The calculation for leg 2 is made directly after leg 1. The initial value $z[T, -g]$ is determined such that the proper length is unchanged. Perrin makes the calculation for leg 3, directly after leg 1. He calculates the initial value $z[2T, -g]$, from the proper length, at the end of leg 2. His argument is that at the end of leg 2, both observers are in the same inertial frame, and they would obtain the same proper length. He then uses the proper length calculated by I. This calculation does not make use of any result obtained from I.

Problems

1. In the case of weak gravity, find h^{00}, h^{0i}, h^{ij} in terms of h_{00}, h_{01}, h_{ij}.

2. For the static case of weak gravity and a spherically symmetric source, the following condition was found:

$$g_{\mu\nu} = \eta_{\mu\nu} + h_{\mu\nu}, \quad |h_{\mu\nu}| \ll 1,$$

$$g^{\mu\nu} = \eta^{\mu\nu} - h^{\mu\nu}, \quad |h^{\mu\nu}| \ll 1,$$

$$h_{00} = -2\Psi_G = 2M'/r,$$

where r is the radial coordinate from the gravitational source mass M'. The result $h_{0i} = h^{0i} = h_{ij} = h^{ij} = 0$ holds. Otherwise, the metric would be direction dependent. In this case, all the h_{ii} have to be equal and nonzero, as no direction is favored,

$$(d\tau)^2_{\text{weak}} = (1 - 2M'/r)(dt)^2 - (1 + h_{ii})[(dx)^2 + (dy)^2 + (dz)^2].$$

Compare $(d\tau)^2_{\text{weak}}$ with the Schwarzschild result for large r, and obtain h_{ii}.

3. In the weak gravity case, GR and Newtonian gravity agree. Here, one can take a stationary state with $\Lambda = U^i = 0$. Let $T_{00} = \rho_M$ be the only nonzero element of $T_{\mu\nu}$, where ρ_M is spherically symmetric. Show that $R = a\rho_M$, where a is the constant in Eq. (5.8). Show $R_{00} = a\rho_M/2 = R^i_{\ 0i0} = -\nabla^2 h_{00}/2$. Thus, $a = 8\pi$.

4. Show that the Einstein field equations, with nonzero cosmological constant, can be written as $R_{\mu\nu} = 8\pi(T_{\mu\nu} - g_{\mu\nu}T^\xi_{\ \xi}/2) + g_{\mu\nu}\Lambda$.

5. Start with Eq. (5.11). Apply Eq. (5.12), and show that Eq. (5.13) is obtained. Then apply Eq. (5.14), and show that Eq. (5.15) is obtained.

6. Repeat the calculation of the Schwarzschild metric, with the cosmological constant Λ included. Find $\exp[2\Phi]$ and $\exp[-2\Delta]$. You should find that the term with Λ is multiplied by r^2. Thus, even though the

data of Chapter 9 yield $\Lambda \approx 5 \times 10^{-54}\,\mathrm{m}^{-2}$, one cannot say that for very large r this term is small. However, on a solar system scale, it is negligible. In this region, weak gravity is satisfied. Find h_{00}, Ψ_G and $\vec{F}_{\text{Newton}}$. Interpret the Λ term.

7. Transform the metric of SR to a frame with coordinates $x^{\mu'}$ where

$$x^0 = [a^{-1} + x^{3'}]\sinh(ax^{0'}), \qquad x^3 = [a^{-1} + x^{3'}]\cosh(ax^{0'}) - a^{-1},$$

$$x^i = x^{i'}, \quad i = i' = 1,2.$$

For small a such that $ax^{0'} = at' \ll 1$, show the transform is to a non-relativistic uniformly accelerating frame. What are the constants of the motion? Calculate $(d\tau)^2$ in the accelerating frame without approximation. Find the constants of the motion. In this frame, clocks at $x^{3'} = 0, h$ are at rest, and measure proper times. What is $d\tau_h/d\tau_0$? Explain the result.

8. Consider the following Schwarzschild-like metric, from the expression for $(d\tau)^2$,

$$(d\tau)^2 = \exp[2\Phi(r)](dt)^2 - (\exp[2\Lambda(r)](dr)^2 + r^2[(d\theta)^2 + \sin^2\theta(d\phi)^2]).$$

Find all the conserved components of a freely falling particle's covariant momentum vector. Show that if the geodesic begins with $x^1 = \theta = \pi/2$ and $P^1 = 0$, these values never change.

9. Repeat the calculations of Problem 8 for the Robertson–Walker metric,

$$(d\tau)^2 = (dt)^2 - Q^2(t)((1 - kr^2)^{-1}(dr)^2 + r^2[(d\theta)^2 + \sin^2\theta(d\phi)^2]).$$

If $k = 0$, $\theta = \pi/2$, and $P^1(0) = 0$, what can be said about $P_3 = P_r$?

10. At earth's surface, electrons and positrons of rest energy 0.511 MeV annihilate at rest into two photons, $e^- + e^+ \to \gamma + \gamma$. One photon makes it to a faraway static, spherically symmetric, compact star. It has mass $M = 1.5M_s$ and surface radial coordinate $\bar{R} = 10$ km. Explain why a single gamma ray in the final state is impossible. At the star's surface, what is the photon energy in MeV? If the decay occurred at the star, what photon energy would be measured at earth?

11. For radial motion, show that the derivatives with respect to τ of the Schwarzschild coordinates are not the energy and momentum per unit rest mass of a particle in a locally inertial frame. Show how they are related to those quantities.

12. For the case $U^i = 0$, calculate the magnitude of the acceleration $a = (\frac{dU_\mu}{d\tau} \frac{U^\mu}{d\tau})^{1/2}$ in the Schwarzschild metric. At the surfaces of the earth and sun, show the result is very close to the expected Newtonian result. Suppose you are just outside a static Schwarzschild black hole where, $2M' = R$ and $r = R + \delta$, $\delta/R \ll 1$. Show that in order to have an acceleration g, R would have to be enormous.

Chapter 6

Classic Solar System Tests
of General Relativity

6.1 Equations of Motion

The metric in the solar system is due mainly to the relativistic mass of the sun $M' = M_s$. The general two-body problem in GR has not been solved analytically, so the sun's mass will be taken as much larger than that of the planet considered, and the gravitational effects of other planets will be neglected. Also the sun's rotation will be neglected. Under these conditions, the metric is that obtained by Schwarzschild. In this chapter, spherical coordinates will be used and $r^p \equiv (r)^p$. The equations of motion involve the C symbols, and from Eqs. (5.16)–(5.18),

$$\exp[-2\Delta] = 1 - 2M'/r, \qquad -2\Delta = \ln[1 - 2M'/r],$$

$$\exp[2\Phi] = 1 - 2M'/r, \qquad 2\Phi = \ln[1 - 2M'/r],$$

$$\Gamma_{rt}^t = \Phi_{,r} = (1 - 2M'/r)^{-1}(M'/r^2),$$

$$\Gamma_{rr}^r = \Delta_{,r} = -(1 - 2M'/r)^{-1}(M'/r^2), \qquad (6.1)$$

$$\Gamma_{tt}^r = \exp[2\Phi]\exp[-2\Delta]\Phi_{,r} = (1 - 2M'/r)(M'/r^2),$$

$$\sin^2\theta\,\Gamma_{\theta\theta}^r = -r\sin^2\theta(1 - 2M'/r) = \Gamma_{\phi\phi}^r.$$

Two constants of the motion are expected for massive particles, since $g_{\mu\nu,0}$ and $g_{\mu\nu,2} = 0$. However, the motion of both photons and massive particles is desired. So the equations of motion (3.25) are written in terms of an affine parameter q,

$$0 = \frac{d^2 x^\mu}{dq^2} + \Gamma^\mu_{\chi\nu} \frac{dx^\chi}{dq} \frac{dx^\nu}{dq}.$$

The value of this parameter will emerge from the solution of the equations.

Problem 3.12 showed that a planar solution $\theta = \pi/2$ is allowed. Thus, for the coordinate $x^2 = \phi$,

$$0 = \frac{d^2\phi}{dq^2} + 2\Gamma^\phi_{r\phi} \frac{dr}{dq} \frac{d\phi}{dq} + 2\Gamma^\phi_{\phi\theta} \frac{d\phi}{dq} \frac{d\theta}{dq}$$

$$= \frac{d^2\phi}{dq^2} + \frac{2}{r} \frac{dr}{dq} \frac{d\phi}{dq}$$

$$= \frac{d^2\phi}{dq^2} \Big/ \frac{d\phi}{dq} + \frac{2}{r} \frac{dr}{dq}$$

$$= \frac{d[\ln(\frac{d\phi}{dq}) + \ln r^2]}{dq}$$

$$= \frac{d[\ln r^2 \frac{d\phi}{dq}]}{dq}, \quad \text{so,}$$

$$J = r^2 \frac{d\phi}{dq}, \qquad d\phi = (J/r^2) dq, \tag{6.2}$$

where J is the constant of the motion due to $g_{\mu\nu,2} = 0$.

For the coordinate $x^0 = t$,

$$0 = \frac{d^2 t}{dq^2} + 2\Gamma^t_{rt} \frac{dr}{dq} \frac{dt}{dq}$$

$$= \frac{d^2 t}{dq^2} + \frac{1}{1 - 2M'/r} \frac{2M'}{r^2} \frac{dr}{dq} \frac{dt}{dq}$$

$$= \frac{d^2 t}{dq^2} \Big/ \frac{dt}{dq} + \frac{1}{1 - 2M'/r} \frac{2M'}{r^2} \frac{dr}{dq}$$

$$= \frac{d(\ln \frac{dt}{dq} + \ln[1 - 2M'/r])}{dq}$$

$$= \frac{d \ln[(1 - 2M'/r) \frac{dt}{dq}]}{dq},$$

$$J' = (1 - 2M'/r)\frac{dt}{dq} \equiv 1,$$

$$\frac{dt}{dq} = \frac{1}{1 - 2M'/r}, \quad dt = \frac{dq}{1 - 2M'/r}. \tag{6.3}$$

To first order in terms of M'/r, $dt \approx dq(1+2M'/r)$. Setting constant $J' = 1$ is a normalization choice. It makes sense because when, $r \to \infty$, $dt = dq$. Since $M'/r \ll 1$, Eq. (6.2) gives $r^2 \frac{d\phi}{dq} \approx r^2 \frac{d\phi}{dt} = J$. This is conservation of the ϕ component of angular momentum per unit mass of the particle moving in the metric.

The above information is inserted into the equation for $x^3 = r$. After multiplication by $\frac{1}{1-2M'/r}\frac{dr}{dq}$, the solution is transparent,

$$0 = \frac{d^2r}{dq^2} + \Gamma^r_{rr}\left(\frac{dr}{dq}\right)^2 + \Gamma^r_{\theta\theta}\left(\frac{d\theta}{dq}\right)^2 + \Gamma^r_{\phi\phi}\left(\frac{d\phi}{dq}\right)^2 + \Gamma^r_{tt}\left(\frac{dt}{dq}\right)^2$$

$$= \frac{d^2r}{dq^2} - \frac{(M'/r^2)\left(\frac{dr}{dq}\right)^2}{1 - 2M'/r}$$

$$+ (1 - 2M'/r)\left[-r\left(\frac{d\phi}{dq}\right)^2 + (M'/r^2)\left(\frac{dt}{dq}\right)^2\right]$$

$$= \frac{d^2r}{dq^2} - \frac{(M'/r^2)(\frac{dr}{dq})^2}{1 - 2M'/r} - \frac{(1 - 2M'/r)J^2}{r^3} + \frac{(M'/r^2)}{1 - 2M'/r}$$

$$= \frac{\frac{d^2r}{dq^2}\frac{dr}{dq}}{1 - 2M'/r} - \frac{(M'/r^2)\left(\frac{dr}{dq}\right)^3}{(1 - 2M'/r)^2} - \frac{J^2\frac{dr}{dq}}{r^3} + \frac{(M'/r^2)\frac{dr}{dq}}{(1 - 2M'/r)^2}$$

$$= \frac{1}{2}\frac{d}{dq}\left[\frac{\left(\frac{dr}{dq}\right)^2}{1 - 2M'/r} + \frac{J^2}{r^2} - \frac{1}{1 - 2M'/r}\right],$$

$$-E' = \frac{\left(\frac{dr}{dq}\right)^2}{1 - 2M'/r} + \frac{J^2}{r^2} - \frac{1}{1 - 2M'/r}, \tag{6.4}$$

where E' is the second constant of the motion, due to $g_{\mu\nu,0} = 0$.

Equation (6.4) allows determination of q and interpretation of E',

$$\left(\frac{dr}{dq}\right)^2 = (1 - 2M'/r)(-E' - (J/r)^2 + (1 - 2M'/r)^{-1}), \tag{6.5}$$

$$(dr)^2 = (dq)^2[-E' - (J/r)^2 + (1 - 2M'/r)^{-1}](1 - 2M'/r),$$

$$(d\tau)^2 = (1 - 2M'/r)(dt)^2 - (1 - 2M'/r)^{-1}(dr)^2 - r^2(d\phi)^2$$

$$= (1 - 2M'/r)^{-1}(dq)^2 - (Jdq/r)^2$$

$$- (dq)^2[-E' - (J/r)^2 + (1 - 2M'/r)^{-1}]$$

$$= E'(dq)^2, \qquad d\tau = E'^{1/2}dq. \tag{6.6}$$

So for photons $E' = 0$, while for massive particles $E' > 0$, and $dq \propto d\tau$. So q is obviously an affine parameter. For a slowly moving particle of rest mass m, with $M' \ll r$, Eqs. (6.3) and (6.5) yield

$$-E' - (J/r)^2 + (1 - 2M'/r)^{-1} = (1 - 2M'/r)^{-3} \left(\frac{dr}{dt}\right)^2$$

$$\approx \left(\frac{dr}{dt}\right)^2,$$

$$-E' \approx \left(\frac{dr}{dt}\right)^2 + (J/r)^2 - (1 + 2M'/r), \tag{6.7}$$

$$(1 - E')/2 = \left[\left(\frac{dr}{dt}\right)^2 + (J/r)^2\right] \Big/ 2 - M'/r,$$

$$m[1 + (1 - E')/2] = RE + KE + PE = E.$$

In general, the quantity E can be interpreted as the total energy or relativistic mass M of the particle. When $E' > 1$, the total energy is less than the rest energy, and may even be negative. That occurs when the particle is in a strong gravitational field.

6.2 Orbit Equations

The orbit equations are obtained from Eqs. (6.2), (6.3), (6.5) and (6.6), e.g.,

$$\frac{d\phi}{dr} = \frac{d\phi}{dq} \Big/ \frac{dr}{dq} = \pm \frac{J}{r^2(1 - 2M'/r)^{1/2}} \frac{1}{([(1 - 2M'/r)^{-1} - E'] - [J/r]^2)^{1/2}},$$

$$D[r] \equiv \frac{1}{([(1 - 2M'/r)^{-1} - E']/J^2 - 1/r^2)^{1/2}}. \tag{6.8}$$

Thus,

$$d\phi = \pm \frac{dr}{r^2 (1 - 2M'/r)^{1/2}} D[r], \tag{6.9}$$

$$dt = \pm \frac{dr}{J(1 - 2M'/r)^{3/2}} D[r], \tag{6.10}$$

$$d\tau = \pm \frac{dr \, E'^{1/2}}{J(1 - 2M'/r)^{1/2}} D[r]. \tag{6.11}$$

It should be noted that t is the time according to a faraway at-rest observer, while τ is the time on a clock attached to the particle. Integration of the above equations will give r as a function of ϕ, t, or τ. The correct sign is determined by the result of the integration. Each integral is an elliptic integral, and could be worked out numerically. However, if $M'/r \ll 1$ everywhere, then to first order in M'/r, the integration can be done analytically. In order to accomplish this, one expands $(1 - 2M'/r)^n = 1 - n(2M'/r) + [n(n-1)/2](2M'/r)^2 + \cdots$, and keeps the lowest order term. In some of the mathematical manipulations below, I have followed Weinberg (1972), and filled in some steps.

6.3 Light Deflection

Solar system tests have shown the correctness of GR, even if equations for the metric, more complicated than Einstein's, cannot be ruled out. Einstein's fame was established by the positive result, of close to the predicted value, for the deflection of light passing near the sun. The first positive result experiment was carried out in 1919, by a British team (Dyson, 1920) led by Eddington.

In the case of light deflection by the sun's gravity, one looks at stars, whose lines of sight, come as close as possible to the edge of the sun's disk. Of course, there must be a complete solar eclipse, as in Fig. 6.1, to make such a measurement. Six months previous to the eclipse, the same stars were viewed. At that time, the sun is no longer between the earth and the stars. In practice, stars closer than twice the sun's radius $2R_s$ cannot be observed. Light from the star initially travels in almost zero gravity, along the line of sight given by angle $\phi(I)$. Its path deflects as it gets close to the sun's surface, where the minimum distance from the sun's center is r_0. The light finally winds up very faraway, along the line of sight given by angle $\phi(F)$. The earth–sun distance is $> 200R_s$, so the metric at earth is

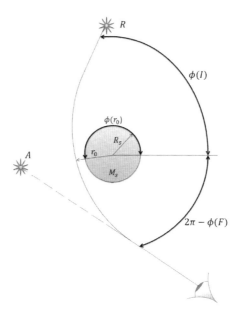

Fig. 6.1   When the sun is in total eclipse, light from stars passing close to the sun's disk appears at apparent position A. When the sun is not between the earth and star, the latter appears at its real position R.

again very close to $\eta_{\mu\nu}$. The angle ϕ is very well measured, and the angular deflection from the initial direction is $\delta\phi = \phi(F) - \phi(I) - \pi$. However, the orbit is symmetric about the line along r_0, the distance of closest approach, so that $\delta\phi = 2(\phi(r_0) - \phi(I)) - \pi$.

For light use Eq. (6.8), with $E' = 0$, and $M'/r_0 \ll 1$. At $r = r_0$, use Eqs. (6.2) and (6.4),

$$\frac{dr}{dq} \propto \frac{dr}{d\phi} = 0,$$

$$J^{-2} = (1 - 2M'/r_0)r_0^{-2}, \tag{6.12}$$

From Eq. (6.9),

$$\delta\phi + \pi = \pm 2 \int_{\infty}^{r_0} dr (1 - 2M'/r)^{-1/2} r^{-2} D[r], \tag{6.13}$$

$$D[r] = \left(J^{-2}(1 - 2M'/r)^{-1} - r^{-2}\right)^{-1/2}$$

$$= \left(r_0^{-2}\frac{1 - 2M'/r_0}{1 - 2M'/r} - r^{-2}\right)^{-1/2}. \tag{6.14}$$

On expanding the terms with M'/r or M'/r_0, and keeping only factors linear in M'/r or M'/r_0, the integral becomes, see Problem 1,

$$\delta\phi + \pi = \pm 2 \int_{\infty}^{r_0} dr \frac{1}{r[(r/r_0)^2 - 1]^{1/2}} \left(1 + \frac{M'}{r_0}\frac{r_0}{r} + \frac{M'r}{r_0^2}\left[\frac{r}{r_0} + 1\right]^{-1}\right).$$

Now change variables such that $u = r/r_0$, $dr = r_0 du$,

$$\delta\phi + \pi = \pm 2 \int_{\infty}^{1} du \frac{1}{u[u^2 - 1]^{1/2}} \left(1 + \frac{M'}{r_0}\left[u^{-1} + \frac{u}{u+1}\right]\right)$$
$$= -2(-\pi/2 - 2M'/r_0) = \pi + 4M'/r_0,$$
$$\delta\phi = 4M'/r_0. \tag{6.15}$$

Since $4M'/R_s = 8.53 \times 10^{-6}$ r(adians) $= 1.75''$, the effect is very small.

The 1919 data were pictures of stars stored on photographic plates. The distances were measured with calipers. Over a period of six months, there is bound to be a change of scale, due to changes in temperature, and the position of the telescope on the ground. The data were compared with the predicted value by calibrating a scale constant S,

$$\delta\phi = (4M'/R_s)(R_s/r_0) + S(r_0/R_s).$$

Stars with $r_0 \gg R_s$ would not change position, and these give the best fit for S. There were two observing stations. At Sobral, an island near the Brazilian coast, seven stars with $r_0/R_s = 2 - 6$ gave $\delta\phi = (1.98 \pm 0.16)''$. At Principe, an island near the coast of Guinea, five stars with $r_0/R_s = 2 - 6$ gave $\delta\phi = (1.61 \pm 0.40)''$. This was sufficient to confirm that light is affected by gravity. The numerical results are in reasonable agreement with the Einstein prediction.

The effect is small, and the accuracy is such that even including more modern optical measurements, metric equations that are more complicated than Einstein's cannot be ruled out. However, radio telescope measurements made at the VLBI facility (Lebach, 1995) have confirmed Einstein's prediction to the 1% level. Further improvements led to an accuracy of 0.02% (Will, 2006). That's what modern instrumentation can do for you.

Due to gravitational deflection of light, the universe offers interesting "illusions," termed gravitational lensing. As illustrated in Fig. 6.2, mass in the form of a galaxy or cluster of galaxies, is located between earth and a distant source. Light rays from the source are deflected by the intervening mass, and follow different paths to earth. The earth observer can then see,

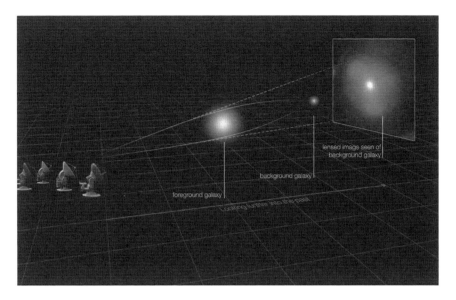

Fig. 6.2 Mass in the form of galaxies and galaxy clusters, between earth and a very faraway source, can deflect light. Multiple images are seen.

in different directions, multiple images of the same source. If the masses line up just right, rings called "Einstein Rings" are observed. Use Google or some other world wide web (WWW) search program to find "gravitational lensing," and you'll find a host of fascinating images.

6.4   Perihelia Advance

The second test is the precession of perhelia of planets close to the sun. This means Mercury for our solar system. The Newtonian orbit for the two-body problem indicates no precession. Mercury's precession is illustrated in Fig. 6.3. While precessing, the orbit will still have maximum, minimum distances from the sun r_+, r_-, where $\frac{dr}{d\phi} = 0$. So Eq. (6.5) is used to obtain J and E' in terms of $r_\pm$,

$$0 = (1 - 2M'/r_+)^{-1} - E' - J^2/r_+^2$$
$$= (1 - 2M'/r_-)^{-1} - E' - J^2/r_-^2,$$
$$J^2 = [r_-^{-2} - r_+^{-2}]^{-1}[(1 - 2M'/r_-)^{-1} - (1 - 2M'/r_+)^{-1}], \qquad (6.16)$$

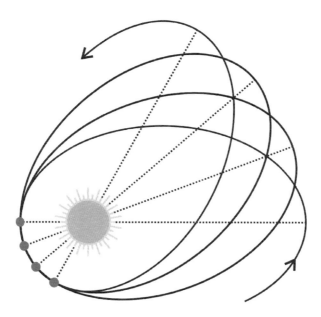

Fig. 6.3   The perihelion of Mercury precessing, as it makes successive orbits about the sun.

$$E' = (1 - 2M'/r_-)^{-1} - J^2 r_-^{-2}$$

$$= (r_+^{-2} - r_-^{-2})^{-1} [r_+^2 (1 - 2M'/r_+)^{-1} - r_-^2 (1 - 2M'/r_-)^{-1}]. \quad (6.17)$$

The integral for the orbit, Eqs. (6.8) and (6.9), requires evaluation of the function $D[r]$, here written in terms of $r_\pm$,

$$f \equiv (D[r])^{-2} + r^{-2} \quad (6.18)$$

$$f = J^{-2}[(1 - 2M'/r)^{-1} - E']$$

$$= (r_+ r_-)^{-2} [(1 - 2M'/r_-)^{-1} - (1 - 2M'/r_+)^{-1}]^{-1}$$

$$\times [r_-^2 ((1 - 2M'/r_-)^{-1} - (1 - 2M'/r)^{-1})$$

$$+ r_+^2 ((1 - 2M'/r)^{-1} - (1 - 2M'/r_+)^{-1})]. \quad (6.19)$$

The above is a positive quantity, as each difference is positive.

Once again the integrand must be expanded in powers of M'/r, and the leading correction kept, in order to do the integral analytically. However, in each difference in the last equation, the leading term is directly proportional to M'. So unless the expansion is carried out to second order, the mass M' will cancel. This expansion is

$$(1 - 2M'/r)^{-1} \approx 1 + 2M'/r + (2M'/r)^2 = 1 + (2M'/r)(1 + 2M'/r).$$

It yields

$$\begin{aligned}
f &= (r_+ r_-)^{-2}[(2M'/r_-)(1 + 2M'/r_-) - (2M'/r_+)(1 + 2M'/r_+)]^{-1} \\
&\quad \times (r_-^2[(2M'/r_-)(1 + 2M'/r_-) - (2M'/r)(1 + 2M'/r)] \\
&\quad + r_+^2[(2M'/r)(1 + 2M'/r) - (2M'/r_+)(1 + 2M'/r_+)]).
\end{aligned} \tag{6.20}$$

Then, one can write

$$(D[r])^{-2} = K(r_-^{-1} - r^{-1})(r^{-1} - r_+^{-1}). \tag{6.21}$$

Note that Eq. (6.20) yields $(D[r_\pm])^{-2} = 0$, and indicates $(D[r])^{-2}$ is a function of r^{-1} and r^{-2}, so that Eq. (6.21) follows. To find the constant K, equate the constant terms in both forms of $(D[r])^{-2}$. This is mathematically equivalent to evaluation at $r = \infty$,

$$\begin{aligned}
K &= (r_+ r_-)^{-1} \\
&\quad \times [2M'/r_+(1 + 2M'/r_+) - 2M'/r_-(1 + 2M'/r_-)]^{-1} \\
&\quad \times (2M'r_-(1 + 2M'/r_-) - 2M'r_+(1 + 2M'/r_+)] \\
&= [r_-(1 + 2M'/r_+) - r_+(1 + 2M'/r_-)]^{-1}(r_- - r_+) \\
&= [1 + 2M'(r_-^{-1} + r_+^{-1})]^{-1} \approx 1 - 2M'(r_-^{-1} + r_+^{-1}), \\
&\quad K^{-1/2} \approx 1 + M'(r_-^{-1} + r_+^{-1}).
\end{aligned} \tag{6.22}$$

When Mercury goes from r_- to r_+, $\delta\phi/2$ is the change in ϕ. As the orbit is symmetric, in one revolution the total change in ϕ is $\delta\phi$,

$$\begin{aligned}
\delta\phi &= \pm 2K^{-1/2} \int_{r_-}^{r_+} \frac{dr}{r^2[(1 - 2M'/r)(r_-^{-1} - r^{-1})(r^{-1} - r_+^{-1})]^{1/2}} \\
&\approx \pm 2K^{-1/2} \int_{r_-}^{r_+} \frac{dr(1 + M'/r)}{r^2[(r_-^{-1} - r^{-1})(r^{-1} - r_+^{-1})]^{1/2}}.
\end{aligned} \tag{6.23}$$

If one makes the change of variable,

$$u = a/r + b, \text{ such that } u[r_\pm] = \pm 1, \qquad (6.24)$$

the integral becomes

$$\delta\phi = \pm 2K^{-1/2} \int_{-1}^{1} du \frac{1 + \frac{M'}{2}[(r_+^{-1} + r_-^{-1}) + (r_+^{-1} - r_-^{-1})u]}{(1 - u^2)^{1/2}}$$

$$= \pi[1 + M'(r_-^{-1} + r_+^{-1})][2 + M'(r_+^{-1} + r_-^{-1})]$$

$$\approx 2\pi + 3\pi M'(r_-^{-1} + r_+^{-1}). \qquad (6.25)$$

If the orbit is closed, the above should equal 2π. So the perihelion has advanced by $3\pi M'(r_-^{-1} + r_+^{-1})$. For the orbit of Mercury, $r_+ = 6.98 \times 10^{10}$ m and $r_- = 4.60 \times 10^{10}$ m. Thus, per revolution, the advance of perihelion is

$$\delta\phi - 2\pi = 3\pi(1.484 \times 10^3)(0.360 \times 10^{-10}) \text{ r} = 0.104''.$$

Mercury's orbital period is 87.96 d, so that in a century it makes 415 revolutions. The cumulative effect is $\delta\phi - 2\pi = 43''$ per century. Current experimental results are in excellent agreement with Einstein's theory. The deviation is less than 1%. The other planets are farther from the sun. They would yield much smaller values than Mercury.

As opposed to light deflection, where light starts and ends in essentially zero gravity, Mercury is always under the influence of gravity. So when it is observed with light, the light is also traveling under that influence. For a single revolution, one would have to worry about such small corrections. However, the effect is cumulative, and for many revolutions this correction is not a worry. Think about measuring the period of a simple pendulum. The small error one might make for one period, is negligible, if the total time for many periods is measured.

Newtonian physics predicts a much larger advance per century, due to other perturbations of $5557''$. These include $5025''$, due to the precession of earth's rotation axis. The observation is $5600''$, and the extra $43''$ agrees with the prediction of GR. See Clemence (1947) for the best determination of the precession. Before GR, Newtonian physics was thought adequate. For the Mercury problem, there was speculation that unseen mass was between the sun and Mercury or the sun was non-spherical, etc. Einstein was not working on GR to solve the Mercury problem, but realized that the theory could be applied to it. His biographers tell us that he was delirious with joy upon finding those $43''$.

6.5   Radar Signal Delay

When Mercury is near superior conjunction, radar from earth is reflected back, with a delay predicted by GR. This effect is also known as Shapiro delay. The situation is shown in Fig. 6.4, where r_0 is the distance of closest approach to the sun. The straight line Newtonian path just grazes the sun's disk. In reality the two paths are very close to coincidence. The tiny deviation is exaggerated for clarity.

As in the case of light deflection $E' = 0$, and at r_0, $\frac{dr}{dq} = \frac{dr}{dt} = 0$. Equations (6.12) and (6.14) give J^2 and $D[r]$, while Eq. (6.10) yields the integrand of the desired integral,

$$J^2 = r_0^2(1 - 2M'/r_0)^{-1},$$

$$dt = \pm\frac{dr(1 - 2M'/r_0)^{1/2}}{r_0(1 - 2M'/r)^{3/2}} \left(r_0^{-2}\frac{(1 - 2M'/r_0)}{(1 - 2M'/r)} - r^{-2}\right)^{-1/2}$$

$$= \pm dr(1 - 2M'/r)^{-1}\left(1 - \frac{1 - 2M'/r}{1 - 2M'/r_0}\left(\frac{r_0}{r}\right)^2\right)^{-1/2}.$$

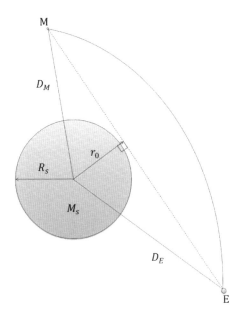

Fig. 6.4   When Mercury is near superior conjunction, radar signals are sent from and reflected back to earth. They travel for a longer time, along the solid GR geodesic, than along the dashed straight, Newtonian path.

In order to avoid elliptic integrals, expand the above, and keep the leading correction,

$$t_{E,M} \approx \pm \int_{r_0}^{D_{E,M}} \frac{r\,dr[1 + M'(2/r + (r_0/r)(r + r_0)^{-1})]}{[r^2 - r_0^2]^{1/2}}, \qquad (6.26)$$

where $D_{E,M}$ is the distance from the center of the sun to the planets. The result is

$$t_{E,M} = M' \left[2\ln \frac{D_{E,M} + [D_{E,M}^2 - r_0^2]^{1/2}}{r_0} + \left(\frac{D_{E,M} - r_0}{D_{E,M} + r_0}\right)^{1/2} \right]$$

$$+ \left[D_{E,M}^2 - r_0^2 \right]^{1/2}. \qquad (6.27)$$

The prediction is compared with radio waves just grazing the sun $r_0 = R_s$. Since the radio waves go to Mercury and are reflected back to earth,

$$t_{\text{total}} = 2(t_E + t_M) \equiv 2(T + T'),$$

$$T = [D_E^2 - R_s^2]^{1/2} + [D_M^2 - R_s^2]^{1/2}, \qquad (6.28)$$

$$T'/M' = 2\ln \frac{(D_E + [D_E^2 - R_s^2]^{1/2})(D_M + [D_M^2 - R_s^2]^{1/2})}{R_s^2}$$

$$+ \left(\frac{D_E - R_s}{D_E + R_s}\right)^{1/2} + \left(\frac{D_M - R_s}{D_M + R_s}\right)^{1/2}$$

$$= 2\ln \frac{D_E(1 + [1 - (R_s/D_E)^2]^{1/2})D_M(1 + [1 - (R_s/D_M)^2]^{1/2})}{R_s^2}$$

$$+ \left(\frac{1 - R_s/D_E}{1 + R_s/D_E}\right)^{1/2} + \left(\frac{1 - R_s/D_M}{1 + R_s/D_M}\right)^{1/2}, \qquad (6.29)$$

where $2T$ is what you would expect if gravity did not deflect radar, while $0 < 2T' \propto M'$ is the lowest order GR correction. As T' is positive, it is said that the radar signals are delayed.

To see the size of the effect, the following parameters are used:

$$D_{E,M} = 0.5(r_+ + r_-)_{E,M} = 1.49 \times 10^{11}\,\text{m}, \ 5.79 \times 10^{10}\,\text{m},$$

$$R_s = 6.96 \times 10^8\,\text{m}, \ M' = 1.48 \times 10^3\,\text{m}, \ \text{thus},$$

$$2T = 4.14 \times 10^{11}\,\text{m} = 1.05 \times 10^3\,\text{s}, \qquad (6.30)$$

$$2T' \approx 4M'\left[1 + \ln 4 \frac{D_E}{R_s}\frac{D_M}{R_s}\right] = 7.2 \times 10^4\,\text{m} = 240\ \mu\text{s}.$$

Since $T'/T = 2.3 \times 10^{-7}$, the delay is a very small effect. Even so, by 1971 the team led by Shapiro achieved a result in agreement with GR at the 5% level (Shapiro, 1971). A similar experiment, using the Cassini space craft, obtained a result in agreement with GR at a <1% level (Berlotti, 2003). It has now been improved to the 0.001% (Will, 2006) level.

A measurement of such exquisite accuracy is fraught with difficulties. Modern atomic clocks can easily measure t_{total} with sufficient accuracy. However, one must subtract from this $2T$, a quantity almost as large. Optical measurements of the distances are nowhere near accurate enough. There are other critical sources of systematic error that must be considered: The solar corona is an effective index of refraction for radar propagation. Reflection at Mercury's surface occurs, not from a single point, but from a large area. Each area element has its own reflection properties, plus orbital and rotational motion. The reflection properties are studied at inferior conjunction. The times of the Doppler-shifted frequency distribution allow one to obtain the reflection time from the closest point to earth. Using a large set of astronomical data, the distances were obtained from GR predictions, in terms of fits for many parameters. See Shapiro (1964) for a description of the original experiment.

Problems

1. Without using the equations of motion, but using $g_{\mu\nu,(0,2)} = 0$, obtain Eqs. (6.9)–(6.11). Do this for photons and particles with rest mass, by appropriate choice of constants.

2. Consider the integrand ($\equiv$ Int) of Eqs. (6.13) and (6.14). Expand the terms involving $M'/r \ll 1$, and keep only terms linear in M'/r. Fill in the steps to obtain the equation following Eq. (6.14).

3. Start with Eqs. (6.16)–(6.18) and fill in the steps leading to Eq. (6.19). Then start with the integral of Eq. (6.23), make the variable change of Eq. (6.24) and fill in the steps leading to Eq. (6.25).

4. Start with the equation above Eq. (6.26) and show how Eq. (6.26) is obtained.

5. Using the Schwarzschild metric, suppose a particle with finite rest mass is launched from the surface of a sphere of area A', in the direction of increasing radial coordinate. The stationary curvature is provided by the sphere's mass M'. Assume $M'/R' \ll 1$, where R' is the radial

coordinate at the sphere's surface. What is the radial coordinate at the surface in terms of A'? Find the escape speed as measured by an observer at the surface of the sphere. Show that this speed is the same as the Newtonian result. What proper distance is traveled by the particle, when it reaches a radial coordinate, where an observer measures that the escape speed is reduced by half?

6. Use the data of and the information from solving Problem 5. Calculate the time on a clock attached to the moving mass, for the trip between the two radial coordinates. Then calculate the time on a clock at rest at $r = \infty$.

7. Use the data of and the information from solving Problem 5. Suppose a mass is released from rest, at a radial coordinate $r = 100R'$. At the surface of the sphere what is the speed? What is the proper distance traveled and the time on a clock attached to the moving mass?

8. The Schwarzschild metric permits circular orbits, where r is constant. For massive particles, show that two such orbits are possible if $12(M'^2/\bar{J}^2) < 1$, where $\bar{J}^2 = J^2/E'$. Show that the larger orbit is stable, but not the smaller. For the latter, what is the smallest possible radial coordinate? Start by showing,

$$1/E' = \left(\frac{dr}{d\tau}\right)^2 + \left(1 - \frac{2M'}{r}\right)\left(1 + \frac{\bar{J}^2}{r^2}\right) \equiv \left(\frac{dr}{d\tau}\right)^2 + f[r].$$

The function $f[r]$ is an effective potential, and radial coordinates of circular orbits exist where $\frac{df[r]}{dr} = 0$.

9. For the conditions of and the information from solving Problem 8 with $r = 10M'$, find J and E'. Determine whether there is another possible circular orbit. If another is possible, find its radial coordinate, and determine which orbit is stable.

10. Use the data of and the information from solving Problems 8 and 9. Find the time on a clock attached to the particle and on a faraway clock at rest, when one revolution is completed. Compare $\frac{d\phi}{dt}$ with the Newtonian value for the same circular orbit.

11. Use the data of and the information from solving Problems 8 and 9. Find the speed $|\vec{v}|$ of the object in the circular orbit, as determined on a clock at rest at the radial coordinate of the orbit. Determine the speed for $r = (10, 4.29, 3)M'$. What can be said because of the speed at $r = 3M'$? Is there a stable circular orbit for light?

12. A neutron star has a mass $M' = 1.5M_s$ and a radius $r_0 = 15$ km. If the star's angular momentum is neglected, the metric in its vicinity is the Schwarzschild metric. If light from faraway just skims its surface, what is the deflection? Use a program like Mathematica to do the numerical integration. Be careful at the limit where $r = r_0$. Compare this result with that obtained by expanding and keeping M'/r terms to lowest order.

Chapter 7

Gravitational Waves

7.1 The Weak Gravity Wave Equation

Einstein's equations predict gravitational waves. The sources of the waves typically involve very strong gravitational effects, for example, the onset of a supernova, or the merging of black holes. However, such sources are usually so faraway that the wave amplitude at earth is extremely small. The wave can be treated in the weak gravity approximation. At this time, it would be beneficial to review the material in Section 5.2 and Problem 5.1. There, in rectangular coordinates, $g_{\mu\nu} = \eta_{\mu\nu} + h_{\mu\nu}$, $|h_{\mu\nu}| \ll 1$, and $|h_{\mu\nu,\xi}| \ll 1$. In deriving the wave equation, some important properties of $h_{\mu\nu}$ and $\bar{h}_{\mu\nu}$, the trace inverse defined below, are of importance,

$$
\begin{aligned}
&h_{\mu\nu} = h_{\nu\mu}, \\
&h^{00} = h_{00}, \ h^{ij} = h_{ij}, \ h^{0i} = -h_{0i}, \\
&h^{\chi}{}_{\nu} = g^{\chi\mu} h_{\mu\nu} \\
&\quad\quad = (\eta^{\chi\mu} - h^{\chi\mu}) h_{\mu\nu} \approx \eta^{\chi\mu} h_{\mu\nu}, \\
&h \equiv h^{\nu}{}_{\nu}, \\
&\bar{h}_{\mu\nu} \equiv h_{\mu\nu} - \eta_{\mu\nu} h/2.
\end{aligned}
\tag{7.1}
$$

These equations lead to

$$
\bar{h}^{\beta}{}_{\nu} = h^{\beta}{}_{\nu} - \delta^{\beta}{}_{\nu} h/2, \ \ \bar{h} = -h.
\tag{7.2}
$$

In order to get to a wave equation, a gauge transformation is needed. Such transforms are encountered in electrodynamics, where the vector potentials $\vec{A}$ and $\vec{A}' = \vec{A} + \vec{\nabla}\Psi$ yield the same magnetic field. However, if $\vec{\nabla} \cdot \vec{A} \neq 0$, then Ψ can be chosen so that $\vec{\nabla} \cdot \vec{A}' = 0$. Here, each coordinate is slightly changed, such that a new metric, also approximately that of SR is obtained. The gauge vectors are functions of the coordinates $\xi^\mu[x^\nu]$. They are taken to be small $|\xi^\mu| \ll 1$ as are their derivatives $|\xi^\mu{}_{,\beta}| \ll 1$. To first order in small quantities,

$$x^{\alpha'} \equiv x^\alpha + \xi^\alpha[x^\beta],$$

$$x^\alpha = x^{\alpha'} - \xi^\alpha[x^\beta] = x^{\alpha'} - \xi^\alpha[x^{\beta'} - \xi^\beta] \approx x^{\alpha'} - \xi^\alpha[x^{\beta'}],$$

$$\xi^\alpha{}_{,\beta} = \xi^\alpha{}_{,\beta'}\, x^{\beta'}{}_{,\beta} = \xi^\alpha{}_{,\beta'}\, (x^\beta + \xi^\beta){}_{,\beta}$$

$$\approx \xi^\alpha{}_{,\beta'},$$

$$\delta^{\alpha'}{}_{\beta'} = x^{\alpha'}{}_{,\beta'} = (x^\alpha + \xi^\alpha){}_{,\beta}\, x^\beta{}_{,\beta'}$$

$$= (\delta^\alpha{}_\beta + \xi^\alpha{}_{,\beta})(x^{\beta'} - \xi^\beta){}_{,\beta'}$$

$$= (\delta^\alpha{}_\beta + \xi^\alpha{}_{,\beta})(1 - \xi^\beta{}_{,\beta'})$$

$$= \delta^\alpha{}_\beta + \xi^\alpha{}_{,\beta} - \delta^\alpha{}_\beta \xi^\beta{}_{,\beta'}$$

$$= \delta^\alpha{}_\beta + \xi^\alpha{}_{,\beta} - \xi^\alpha{}_{,\beta'}$$

$$= \delta^\alpha{}_\beta = x^\alpha{}_{,\beta}, \text{ therefore } \eta_{\mu'\nu'} = \eta_{\mu\nu},$$

$$x^\alpha{}_{,\beta'} = x^{\alpha'}{}_{,\beta'} - \xi^\alpha{}_{,\beta'} = \delta^\alpha{}_\beta - \xi^\alpha{}_{,\beta},$$

$$x^{\alpha'}{}_{,\beta} = x^\alpha{}_{,\beta} + \xi^\alpha{}_{,\beta} = \delta^\alpha{}_\beta + \xi^\alpha{}_{,\beta},$$

$$g_{\alpha'\beta'} = \eta_{\alpha'\beta'} + h_{\alpha'\beta'} = \eta_{\alpha\beta} + h_{\alpha'\beta'}. \tag{7.3}$$

These equations lead to

$$h_{\alpha\beta} = h_{\alpha'\beta'} + \xi_{\alpha,\beta} + \xi_{\beta,\alpha},$$
$$h_{\alpha'\beta'} = h_{\alpha\beta} - (\xi_{\alpha,\beta} + \xi_{\beta,\alpha}). \tag{7.4}$$

The gist of all this is, that $h_{\mu\nu}$ is not a unique tensor, but it remains small when changed by a gauge transformation. The new form, obtained from $g_{\mu'\nu'} = x^\xi{}_{,\mu'}\, x^\chi{}_{,\nu'} g_{\xi\chi}$, may be more advantageous to solve a particular problem.

To first order, the Einstein tensor becomes a wave equation. The calculation begins with the curvature tensor,

$$R_{\alpha\beta\mu\nu} = g_{\alpha\chi}R^{\chi}{}_{\beta\mu\nu} = g_{\alpha\chi}[\Gamma^{\chi}{}_{\beta\nu,\mu} - \Gamma^{\chi}{}_{\beta\mu,\nu} + \Gamma^{\chi}{}_{\gamma\mu}\Gamma^{\gamma}{}_{\beta\nu} - \Gamma^{\chi}{}_{\gamma\nu}\Gamma^{\gamma}{}_{\beta\mu}],$$

$$\Gamma^{\chi}{}_{\beta\mu} = g^{\sigma\chi}[g_{\beta\sigma,\mu} + g_{\mu\sigma,\beta} - g_{\beta\mu,\sigma}]/2 = \eta^{\sigma\chi}[h_{\beta\sigma,\mu} + h_{\mu\sigma,\beta} - h_{\beta\mu,\sigma}]/2.$$

The $\Gamma\Gamma$ terms in the curvature tensor are at least second order in $h_{\mu\nu,\xi}$ and can be neglected. The derivative terms become

$$\Gamma^{\chi}{}_{\beta\mu,\nu} = \eta^{\sigma\chi}[h_{\beta\sigma,\mu,\nu} + h_{\mu\sigma,\beta,\nu} - h_{\beta\mu,\sigma,\nu}]/2,$$

$$\Gamma^{\chi}{}_{\beta\nu,\mu} = \eta^{\sigma\chi}[h_{\beta\sigma,\nu,\mu} + h_{\nu\sigma,\beta,\mu} - h_{\beta\nu,\sigma,\mu}]/2$$

$$= \eta^{\sigma\chi}[h_{\beta\sigma,\mu,\nu} + h_{\nu\sigma,\beta,\mu} - h_{\beta\nu,\sigma,\mu}]/2.$$

The curvature tensor is

$$R_{\alpha\beta\mu\nu} = \eta_{\alpha\chi}\eta^{\sigma\chi}[h_{\nu\sigma,\beta,\mu} - h_{\beta\nu,\sigma,\mu} - h_{\mu\sigma,\beta,\nu} + h_{\beta\mu,\sigma,\nu}]/2$$

$$= [h_{\nu\alpha,\beta,\mu} - h_{\beta\nu,\alpha,\mu} - h_{\mu\alpha,\beta,\nu} + h_{\beta\mu,\alpha,\nu}]/2.$$

Using the metric, the Ricci tensor can be calculated,

$$R_{\beta\nu} = g^{\alpha\mu}R_{\alpha\beta\mu\nu} = \eta^{\alpha\mu}R_{\alpha\beta\mu\nu}$$

$$= [h_{\nu}{}^{\mu},{}_{\beta,\mu} - \eta^{\alpha\mu}h_{\beta\nu,\alpha,\mu} - h_{\mu}{}^{\mu},{}_{\beta,\nu} + h_{\beta}{}^{\alpha},{}_{\alpha,\nu}]/2$$

$$= [h_{\nu}{}^{\mu},{}_{\beta,\mu} - \eta^{\alpha\mu}h_{\beta\nu,\alpha,\mu} - h,{}_{\beta,\nu} + h_{\beta}{}^{\alpha},{}_{\alpha,\nu}]/2. \tag{7.5}$$

Note, second term in the above equation is

$$-\left(\nabla^2 - \frac{\partial^2}{\partial t^2}\right)h_{\beta\nu} \equiv -\Box h_{\beta\nu}.$$

So one gets an inkling of how the wave equation comes about.

The Ricci scalar and the Einstein tensor are now evaluated. Renaming some summed over indexes helps carry out the calculation,

$$R = g^{\beta\nu}R_{\beta\nu} = \eta^{\beta\nu}R_{\beta\nu}$$

$$= \eta^{\beta\nu}[h_{\nu}{}^{\mu},{}_{\beta,\mu} - \eta^{\alpha\mu}h_{\beta\nu,\alpha,\mu} - h,{}_{\beta,\nu} + h_{\beta}{}^{\alpha},{}_{\alpha,\nu}]/2$$

$$= [h^{\beta\mu},{}_{\mu,\beta} - \eta^{\alpha\mu}h^{\beta}{}_{\beta,\mu,\alpha} - \eta^{\beta\nu}h,{}_{\beta,\nu} + h^{\nu\alpha},{}_{\alpha,\nu}]/2$$

$$= h^{\mu\alpha},{}_{\mu,\alpha} - \eta^{\alpha\mu}h,{}_{\mu,\alpha}.$$

This yields the Einstein tensor,

$$G_{\beta\nu} = R_{\beta\nu} - g_{\beta\nu}R/2 \approx R_{\beta\nu} - \eta_{\beta\nu}R/2$$
$$= [h_\nu{}^\mu{}_{,\beta}{}_{,\mu} - \eta^{\alpha\mu}h_{\beta\nu,\alpha}{}_{,\mu} - h_{,\beta}{}_{,\nu} + h_\beta{}^\alpha{}_{,\alpha}{}_{,\nu}$$
$$- \eta_{\beta\nu}(h^{\mu\alpha}{}_{,\mu}{}_{,\alpha} - \eta^{\alpha\mu}h_{,\mu}{}_{,\alpha})]/2.$$

If the above is written in terms of the variable $\bar{h}$, the trace inverse instead of h, and a gauge transform is applied, out pops the following wave equation:

$$G_{\beta\nu} = (1/2)[(\bar{h}_\nu{}^\mu - \delta_\nu{}^\mu\bar{h}/2)_{,\beta}{}_{,\mu} - \eta^{\alpha\mu}(\bar{h}_{\beta\nu} - \eta_{\beta\nu}\bar{h}/2)_{,\alpha}{}_{,\mu}$$
$$+ \bar{h}_{,\beta}{}_{,\nu} + (\bar{h}_\beta{}^\alpha - \delta_\beta{}^\alpha\bar{h}/2)_{,\alpha}{}_{,\nu}$$
$$- \eta_{\beta\nu}((\bar{h}^{\mu\alpha} - \eta^{\mu\alpha}\bar{h}/2)_{,\mu}{}_{,\alpha} + \eta^{\alpha\mu}\bar{h}_{,\mu}{}_{,\alpha})]$$
$$= (1/2)(\bar{h}_\nu{}^\mu{}_{,\beta}{}_{,\mu} - \eta^{\alpha\mu}\bar{h}_{\beta\nu,\alpha}{}_{,\mu} + \bar{h}_\beta{}^\alpha{}_{,\alpha}{}_{,\nu} - \eta_{\beta\nu}\bar{h}^{\mu\alpha}{}_{,\mu}{}_{,\alpha}$$
$$- [\delta_\nu{}^\mu\bar{h}_{,\beta}{}_{,\mu} - 2\bar{h}_{,\beta}{}_{,\nu} + \delta_\beta{}^\alpha\bar{h}_{,\alpha}{}_{,\nu}$$
$$- (\eta_{\beta\nu}\eta^{\alpha\mu}\bar{h}_{,\alpha}{}_{,\mu} + \eta_{\beta\nu}\eta^{\mu\alpha}\bar{h}_{,\mu}{}_{,\alpha} - 2\eta_{\beta\nu}\eta^{\mu\alpha}\bar{h}_{,\mu}{}_{,\alpha})]/2)$$
$$= [\bar{h}_\nu{}^\mu{}_{,\beta}{}_{,\mu} - \eta^{\alpha\mu}\bar{h}_{\beta\nu,\alpha}{}_{,\mu} + \bar{h}_\beta{}^\alpha{}_{,\alpha}{}_{,\nu} - \eta_{\beta\nu}\bar{h}^{\mu\alpha}{}_{,\mu}{}_{,\alpha}]/2$$
$$= [(\bar{h}_\nu{}^\mu{}_{,\mu})_{,\beta} - \eta^{\alpha\mu}\bar{h}_{\beta\nu,\alpha}{}_{,\mu} + (\bar{h}_\beta{}^\alpha{}_{,\alpha})_{,\nu} - \eta_{\beta\nu}\eta^{\mu\xi}(\bar{h}_\xi{}^\alpha{}_{,\alpha})_{,\mu}]/2.$$

The Lorentz gauge, in analogy with electrodynamics, is $\bar{h}_\sigma{}^\chi{}_{,\chi} = 0$. Then the above equation simplifies to a single term. There are four equations $\bar{h}_\sigma{}^\chi{}_{,\chi} = 0$, one for each σ. There are four free gauge functions ξ^σ. So it is possible to find a gauge, using Eq. (7.4), for which the simplification holds. The Einstein tensor becomes

$$2G_{\beta\nu} = 16\pi T_{\beta\nu} = -\eta^{\alpha\mu}\bar{h}_{\beta\nu,\alpha}{}_{,\mu} = -\Box\bar{h}_{\beta\nu},$$
$$\Box\bar{h}_{\beta\nu} = -16\pi T_{\beta\nu}, \quad \Box\bar{h}^{\beta\nu} = -16\pi T^{\beta\nu}. \tag{7.6}$$

This is a wave equation with a source proportional to the energy–momentum tensor. The wave propagates with the speed of light and all frequencies are allowed.

7.2 Plane Waves

In empty space there is no source, so plane wave solutions are possible. With enough plane waves of different wave vectors and accompanying amplitudes, any wave shape can be accommodated by superposition. In the case of a single wave vector k_χ with amplitude $A_{\mu\nu}$, a complex constant, the wave function in rectangular coordinates is the real part of $\bar{h}_{\mu\nu} = A_{\mu\nu}\exp(ik_\chi x^\chi)$. The phase factor is an invariant. It is easily shown that

$$\bar{h}_{\mu\nu,\beta} = ik_\beta\bar{h}_{\mu\nu}, \tag{7.7}$$

$$\Box\bar{h}_{\mu\nu} = -(k_\beta k^\beta)\bar{h}_{\mu\nu} = 0, \quad k_\beta k^\beta = 0. \tag{7.8}$$

As with an electromagnetic wave, the relation between frequency $k^0 \equiv \omega$ and wave 3-vector $\vec{k}$ is identical. In free space, there is no dispersion, so the phase and group velocities are unity. The direction of $\vec{k}$ is the direction of wave travel. The gauge condition forced $\bar{h}_\mu{}^\nu{}_{,\nu} = 0$. Thus,

$$k_\nu A_\mu{}^\nu = 0. \tag{7.9}$$

This is another restriction, an orthogonality restriction on $A_\mu{}^\nu$.

A more useful solution can be obtained, by again applying a gauge transformation with vector ξ_α. The vector satisfies $\Box\xi_\alpha = 0$. It can produce a solution ${}^{TT}\bar{h}_{\mu\nu}$, with amplitude $\bar{A}_{\mu\nu}$, that is traceless ${}^{TT}\bar{h}_\mu{}^\mu = \bar{A}_\mu{}^\mu = 0$. Using $\xi_\mu = B_\mu\exp(ik_\chi r^\chi)$ and results from Problem 2,

$$\bar{h}_{\mu'\nu'} \equiv {}^{TT}\bar{h}_{\mu\nu} = \bar{h}_{\mu\nu} - \xi_{\mu,\nu} - \xi_{\nu,\mu} + \eta_{\mu\nu}\xi^\chi{}_{,\chi},$$

$$\bar{A}_{\mu\nu} = A_{\mu\nu} - i(B_\mu k_\nu + k_\mu B_\nu - \eta_{\mu\nu}B^\chi k_\chi), \tag{7.10}$$

$$\bar{A}_\mu{}^\alpha = A_\mu{}^\alpha - i(B_\mu k^\alpha + k_\mu B^\alpha - \delta_\mu{}^\alpha B^\chi k_\chi),$$

$$0 = \bar{A}_\mu{}^\mu = A_\mu{}^\mu - i(B_\mu k^\mu + k_\mu B^\mu - 4B^\chi k_\chi) \tag{7.11}$$

$$= A_\mu{}^\mu + 2iB_\mu k^\mu = A_\mu{}^\mu + 2iB^\mu k_\mu, \quad iB^\mu k_\mu = -A_\mu{}^\mu/2.$$

Thus, if $A_\mu{}^\mu \neq 0$, one can choose a B_μ to make $\bar{A}_\mu{}^\mu = 0 = {}^{TT}\bar{h}_\mu{}^\mu$. This gives one requirement for the four B_μ. The other three can be used to impose the condition $\bar{A}_{\mu\nu}U^\nu = 0$ because from Eqs. (7.8)–(7.10),

$$0 = k^\mu\bar{A}_{\mu\nu} = k^\mu A_{\mu\nu} - i(B_\mu k^\mu k_\nu + B_\nu k^\mu k_\mu - k_\nu B^\chi k_\chi) = k^\mu A_{\mu\nu}$$

$$= (\bar{A}_{\mu\nu}k^\mu)U^\nu = k^\mu(\bar{A}_{\mu\nu}U^\nu). \tag{7.12}$$

The remaining three requirements on the four B_μ can force,

$$\bar{A}_{j\nu}U^\nu = 0. \tag{7.13}$$

Then Eqs. (7.12) and (7.13) force,

$$\bar{A}_{0\nu}U^{\nu} = 0. \tag{7.14}$$

In the language of the subject, the superscript TT stands for transverse traceless gauge. In order to see how it is transverse, let the wave be moving in the $z = 3$-direction. Let $U^{\nu} = \delta^{\nu}{}_{0}$,

$$\begin{aligned}
0 = U^{\nu}\bar{A}_{\mu\nu} &= \delta^{\nu}{}_{0}\bar{A}_{\mu\nu} = \bar{A}_{\mu 0} = \bar{A}^{\mu 0} \\
&= k_{\nu}\bar{A}^{\mu\nu} = \omega(-\bar{A}^{\mu 0} + \bar{A}^{\mu 3}) = \omega\bar{A}^{\mu 3}.
\end{aligned}$$

In this case only, $\bar{A}_{(11),(22),(12),(21)}$ are nonzero, and the symmetry and traceless conditions yield

$$\begin{aligned}
0 &= \bar{A}_{\mu}{}^{\mu} = \bar{A}_{1}{}^{1} + \bar{A}_{2}{}^{2}, \\
\bar{A}_{1}{}^{1} &= -\bar{A}_{2}{}^{2}, \quad \bar{A}_{11} = -\bar{A}_{22}, \quad \bar{A}_{12} = \bar{A}_{21}.
\end{aligned} \tag{7.15}$$

In this gauge, the following conditions also hold:

$$\begin{aligned}
-(^{TT}h) &= {}^{TT}\bar{h} = {}^{TT}\bar{h}_{\mu}{}^{\mu} = 0, \\
{}^{TT}\bar{h}_{\mu\nu} &= {}^{TT}h_{\mu\nu} - \eta_{\mu\nu}(^{TT}h)/2 = {}^{TT}h_{\mu\nu}.
\end{aligned} \tag{7.16}$$

7.3   The Graviton

For electromagnetic waves, the wave function of the vector field A_{μ} can describe all of the physics. When this field is quantized, the quanta are photons with spin $s = 1$. In quantum electrodynamics, the interactions to lowest order are the exchange of virtual photons. In GR, the wave function of the field describing the physics is a tensor of rank 2 $\bar{h}_{\mu\nu}$. Thus, a quantum theory of gravity has a exchange particle of spin $s = 2$, with zero rest mass, called the graviton.

    A transparent way to see this is to consider what happens to a transverse electromagnetic or transverse, traceless gravitational plane wave amplitude, under rotation. If the plane wave is traveling in the 3-direction, the only nonzero amplitudes are A_{j} for the electromagnetic wave and $\bar{A}_{jk}$ for the gravitational wave. Here $(j,k) \neq 3$. One can rotate these wave functions by angle ϕ about the axis along the propagation direction, using the information in Fig. 1.1. Another set of rectangular axes, where basis and unit vectors are the same, is obtained. The nonzero elements of the rotation matrix $x^{i}{}_{,j'}$ are: $R^{1}{}_{1'} = R^{2}{}_{2'} = \cos\phi$, $R^{1}{}_{2'} = -R^{2}{}_{1'} = \sin\phi$, $R^{3}{}_{3'} = 1$.

Using the rotation $A_{j'} = R^k_{j'} A_k$, the electromagnetic amplitudes become

$$A_{1'} = R^1_{1'} A_1 + R^2_{1'} A_2 = \cos\phi A_1 - \sin\phi A_2,$$
$$A_{2'} = R^1_{2'} A_1 + R^2_{2'} A_2 = \sin\phi A_1 + \cos\phi A_2.$$

These equations yield

$$A_{1'} \pm iA_{2'} = (\cos\phi \pm i\sin\phi)A_1 + (-\sin\phi \pm i\cos\phi)A_2$$
$$= \exp(\pm i\phi)A_1 \pm (\cos\phi \pm i\sin\phi)iA_2 = \exp(\pm i\phi)[A_1 \pm iA_2].$$

Using the rotation $\bar{A}_{j'k'} = R^l_{j'} R^n_{k'} \bar{A}_{ln}$ and Eq. (7.15), the gravitational amplitudes become

$$\bar{A}_{1'1'} = R^1_{1'} R^1_{1'} \bar{A}_{11} + R^1_{1'} R^2_{1'} \bar{A}_{12} + R^2_{1'} R^1_{1'} \bar{A}_{21} + R^2_{1'} R^2_{1'} \bar{A}_{22}$$
$$= \cos^2\phi \bar{A}_{11} - 2\sin\phi\cos\phi \bar{A}_{12} - \sin^2\phi \bar{A}_{11}$$
$$= \cos 2\phi \bar{A}_{11} - \sin 2\phi \bar{A}_{12},$$
$$\bar{A}_{2'2'} = R^1_{2'} R^1_{2'} \bar{A}_{11} + R^1_{2'} R^2_{2'} \bar{A}_{12} + R^2_{2'} R^1_{2'} \bar{A}_{21} + R^2_{2'} R^2_{2'} \bar{A}_{22}$$
$$= \sin^2\phi \bar{A}_{11} + 2\sin\phi\cos\phi \bar{A}_{12} - \cos^2\phi \bar{A}_{11}$$
$$= -\cos 2\phi \bar{A}_{11} + \sin 2\phi \bar{A}_{12},$$
$$\bar{A}_{1'2'} = R^1_{1'} R^1_{2'} \bar{A}_{11} + R^1_{1'} R^2_{2'} \bar{A}_{12} + R^2_{1'} R^1_{2'} \bar{A}_{21} + R^2_{1'} R^2_{2'} \bar{A}_{22}$$
$$= \sin\phi\cos\phi \bar{A}_{11} + (\cos^2\phi - \sin^2\phi)\bar{A}_{12} + \sin\phi\cos\phi \bar{A}_{11}$$
$$= \sin 2\phi \bar{A}_{11} + \cos 2\phi \bar{A}_{12}.$$

These equations yield

$$\bar{A}_{1'1'} \pm i\bar{A}_{1'2'} = -\bar{A}_{2'2'} \pm i\bar{A}_{1'2'}$$
$$= \exp(\pm 2i\phi)(\bar{A}_{11} \pm i\bar{A}_{12})$$
$$= \exp(\pm 2i\phi)(-\bar{A}_{22} \pm i\bar{A}_{12}).$$

In quantum mechanics, a plane wave is an eigenfunction of linear momentum. If under rotation by angle ϕ about the direction of propagation, the eigenfunction is transformed $\Psi \to \exp(in\phi)\Psi$, then it is also an eigenfunction of helicity $h = n$. In our case, $n = \pm 1$ for electrodynamics and $n = \pm 2$ for gravitation. When the fields are quantized, it leads to spin

$s = 1$ photons, with only the projections $s_z = \pm 1$ along the direction of propagation. For gravity, it leads to spin $s = 2$ gravitons, with only the projections $s_z = \pm 2$. Since string theory naturally allows for a spin $s = 2$ particle, a great many talented theorists study string theory. They hope to unify gravity and the other interactions.

7.4 Gravity Wave Detection: LIGO Experiment

In order to see how the wave affects free particles, consider a particle initially at rest in a Lorentz frame. The velocity components are $U^0 = 1$ and $U^i = 0$. As previously shown, in the TT gauge this velocity forced $^{TT}h_{\mu 0} = 0$. The following equations of motion for the particle's coordinates are obtained

$$0 = \frac{dU^\mu}{d\tau} + \Gamma^\mu_{\nu\chi}U^\nu U^\chi$$

$$\frac{dU^\mu}{d\tau}[x = 0] = -\Gamma^\mu_{00}$$

$$= -\eta^{\mu\sigma}(^{TT}h_{\sigma 0,0} + {}^{TT}h_{0\sigma,0} - {}^{TT}h_{00,\sigma})/2 = 0.$$

So initially the acceleration vanishes. This means that the particle will be at rest at a later time, and by the same argument, the acceleration will be zero at a later time. Thus, the particle can remain at its coordinate in this gauge. This has no invariant geometrical meaning. However, suppose particles at $x = 0, \epsilon$ experience a wave. The proper distance between them changes to

$$l = \int_0^\epsilon (g_{xx})^{1/2}dx \approx \epsilon(g_{xx}[x = 0])^{1/2}$$

$$= \epsilon(1 + {}^{TT}h_{xx}[x = 0])^{1/2}$$

$$\approx \epsilon(1 + {}^{TT}h_{xx}[x = 0]/2). \qquad (7.17)$$

Thus, the wave can be detected because the proper distance between two objects will wiggle in its presence.

Observations of changing pulsing period, of a pulsar in a binary neutron star system, indicate that energy is being carried away. This is an indirect detection of a wave, one that is continuously generated. This system is described in a later section. Calculation indicated that waves from sufficiently close, violent, short-lived, astronomical events could be detectable on earth. Waves from such events have been observed by the

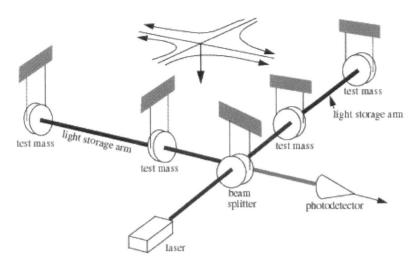

Fig. 7.1   One of two LIGO interferometers. See text for description.

LIGO (2015) experiment via construction of identical interferometers in Washington state and Louisiana.

In Fig. 7.1, one interferometer is illustrated. A laser beam is split and travels down the arms of a long "L." The closest mirrors to the beam splitter are slightly transmitting. The splitter and mirrors are hung in vacuum-like pendulums, so that they are essentially free particles. The light travels back and forth about 100 times in the 4 km long light storage arms. The light from the arms interferes destructively, so that no light gets to the photo-detector. The light heads back towards the laser, where it is recirculated by a power recycling mirror. The latter is not shown in the figure.

If there was a gravitational wave of sufficient power, it would alter a huge region of space. The laser beams wouldn't return to the beam-splitter out of phase. A signal would appear in the photo-detector. This would happen in both interferometers with a separation time of 0.007 s, the light travel time between them. Such a coincidence greatly limits the background noise. The interferometers are updated all the time, and their signal-to-noise resolution in the desirable frequency range has improved, so that $h \approx 10^{-22}$ is close to detectable. That's quite an achievement. However, many expected signals are still in the noise. In order to confirm a wave detection, a coincidence of like signals in the interferometers, plus knowledge of what the signal should look like, is required. The latter is where theoreticians make their

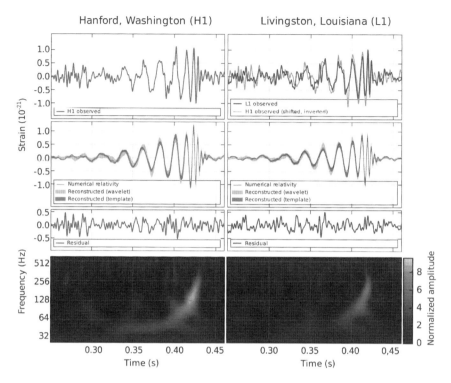

Fig. 7.2 LIGO evidence for the first gravitational wave detection. See text for description.

presence felt. For example, the signals from coalescing black holes, or from supernova, yield very distinctive time structures.

In February, 2016, close to when the first edition of this book went to press, LIGO announced observation of a robust signal. Its time structure is shown in Fig. 7.2. The amplitude $h \approx 10^{-21}$ was due to the merging of two black holes. The event took place about 1 Bly away, and each black hole had equivalent mass $M > 30M_s$. In $<0.5\,$s they merged into a single black hole, and about $4M_s$ was radiated away. This was more power than that of the visible universe. Since then, more such events have been detected. One was seen by LIGO and two newly constructed international detectors, for example, Virgo. This can pinpoint the direction of events by triangulation. Another event, the merger of two neutron stars, was also seen electromagnetically, making it easier to locate the source. A new age of astronomy, of Galilean significance, has begun!

7.5 Wave Equation Solution With Sources

If there are no sources, plane wave solutions are allowed. However, without knowledge of the source, the wave amplitudes and the power carried by the wave, cannot be calculated. When scientists build a detector-like LIGO, they know the minimum signals that can be observed. Before many hundreds of millions of dollars are spent, one would like to be confident, that the sought for gravitational waves, carry enough power to be detected. Thus solution of the wave equation with sources is a necessary part of the program. Such a signal is distinctive from the background noise, even though the two may have the same power level.

Most readers have solved the wave equation with sources in electrodynamics courses. They may skip this section without loss. It is included for completeness, and as a handy reference. The method of Fourier analysis is used to solve Eq. (7.6). This analysis makes use of the following theorem concerning the Dirac delta function:

$$\delta(t' - t) = \frac{1}{2\pi} \int_{-\infty}^{\infty} d\omega \exp[i\omega(t' - t)].$$ (7.18)

One sees that Eq. (7.6) looks like,

$$\Box\Psi[\vec{r}, t] = \left(\nabla^2 - \frac{\partial^2}{\partial t^2}\right) \Psi[\vec{r}, t] = -4\pi S[\vec{r}, t],$$

where $\Psi = \bar{h}^{\mu\nu}$ is the wave function and $S = 4T^{\mu\nu}$ is the source. Using Fourier transforms, one can write

$$\Psi[\vec{r}, t] = \frac{1}{2\pi} \int_{-\infty}^{\infty} d\omega \Psi[\vec{r}, \omega] \exp[-i\omega t],$$

where

$$\Psi[\vec{r}, \omega] = \int_{-\infty}^{\infty} dt' \Psi[\vec{r}, t'] \exp[i\omega t'],$$

since

$$\Psi[\vec{r},t] = \frac{1}{2\pi} \int_{-\infty}^{\infty} d\omega \int_{-\infty}^{\infty} dt' \Psi[\vec{r},t'] \exp[i\omega(t'-t)]$$

$$= \int_{-\infty}^{\infty} dt' \Psi[\vec{r},t'] \frac{1}{2\pi} \int_{-\infty}^{\infty} d\omega \exp[i\omega(t'-t)])$$

$$= \int_{-\infty}^{\infty} dt' \Psi[\vec{r},t'] \delta(t'-t) = \Psi[\vec{r},t].$$

The formal mathematical solution to the wave equation can be expressed as an integral over all space and time,

$$\Psi[\vec{r},t] \equiv \int_{-\infty}^{\infty} dt' \int_{\infty} dV' S[\vec{r}\,',t'] G[\vec{r},t,\vec{r}\,',t'],$$

where

$$\Box \Psi[\vec{r},t] = -4\pi S[\vec{r},t]$$

$$= \int_{-\infty}^{\infty} dt' \int_{\infty} dV' S[\vec{r}\,',t'] \Box G[\vec{r},t,\vec{r}\,',t'],$$

$$\Box G[\vec{r},t,\vec{r}\,',t'] = -4\pi\delta(t'-t)\delta(\vec{r}\,'-\vec{r}) \equiv -4\pi\delta(t'-t)\delta[\vec{R}]$$

$$= -4\pi\delta[\vec{R}]\frac{1}{2\pi}\int_{-\infty}^{\infty} d\omega \exp[i\omega(t'-t)].$$

The function G is called a Green function. It is easily solved for using Fourier transforms,

$$G[\vec{r},t,\vec{r}\,',t'] = \frac{1}{2\pi}\int_{-\infty}^{\infty} d\omega\, G[\vec{r},\omega,\vec{r}\,',t']\exp[-i\omega t],$$

$$\Box G[\vec{r},t,\vec{r}\,',t'] = \frac{1}{2\pi}\int_{-\infty}^{\infty} d\omega \Box (G[\vec{r},\omega,\vec{r}\,',t']\exp[-i\omega t]),$$

$$\Box(G[\vec{r},\omega,\vec{r}\,',t']\exp[-i\omega t]) = -4\pi\delta[\vec{R}]\exp[i\omega(t'-t)],$$

$$(\nabla^2 + \omega^2)G[\vec{r},\omega,\vec{r}\,',t'] = -4\pi\delta[\vec{R}]\exp[i\omega t'],$$

$$G[\vec{r},\omega,\vec{r}\,',t'] \equiv g[R,\omega]\exp[i\omega t'].$$

The last form results because we are in unbounded space, and the source for $G[\vec{r},\omega,\vec{r}\,',t']$ is only nonzero at $\vec{R} = 0$. Thus the spatial dependence can only be radial, a function of R. When $R \neq 0$, the differential equation to solve is

$$0 = (\nabla^2 + \omega^2)g[R,\omega] = (\nabla_R^2 + \omega^2)g[R,\omega]$$

$$= R^{-1}(Rg[R,\omega])_{,R\,,R} + \omega^2 g[R,\omega]$$

$$= (Rg[R,\omega])_{,R\,,R} + \omega^2 (Rg[R,\omega]),$$

$$Rg[R,\omega] = \exp[\pm i\omega R],$$

$$RG[\vec{r},\omega,\vec{r}\,',t'] = \exp[i(\pm\omega R + \omega t')] = \exp[i\omega(\pm R + t')],$$

$$G[\vec{r},\omega,\vec{r}\,',t'] = \exp[i\omega(\pm R + t')]/R,$$

$$G[\vec{r},t,\vec{r}\,',t'] = \frac{1}{2\pi}\int_{-\infty}^{\infty} d\omega\, \frac{\exp[i\omega(\pm R + t' - t)]}{R}$$

$$= |\vec{r} - \vec{r}\,'|^{-1}\delta[t' - (t \mp (|\vec{r} - \vec{r}\,'|))],$$

$$\bar{h}^{\mu\nu}[\vec{r},t] = \int_{\infty} dV' \int_{-\infty}^{\infty} dt'\, S[\vec{r}\,',t'] \frac{\delta[t' - (t \mp (|\vec{r} - \vec{r}\,'|))]}{|\vec{r} - \vec{r}\,'|} \qquad (7.19)$$

$$= \int_{\infty} dV' \frac{S[\vec{r}\,',t - |\vec{r} - \vec{r}\,'|]}{|\vec{r} - \vec{r}\,'|}$$

$$= 4\int_{\infty} dV' \frac{T^{\mu\nu}[\vec{r}\,',t - |\vec{r} - \vec{r}\,'|]}{|\vec{r} - \vec{r}\,'|}.$$

In the last form you note that the minus sign is chosen. Then the wave function at $\vec{r},t$ depends on the source at $\vec{r}\,'$ and at an earlier time. Earlier by $|\vec{r} - \vec{r}\,'|$ because $c = 1$. This is physical as the wave must travel from the source to the observation point. If the plus sign was chosen, then what happens later than t at the source would dictate the wave function at $\vec{r},t$. That is why the solution is called the retarded time solution. Whatever the t' dependence of the source, just substitute $t - |\vec{r} - \vec{r}\,'|$ for t'.

7.6 The Energy–Momentum Tensor

To consider the energy–momentum tensor, it is easiest to begin in an SR frame. Consider a system of n free particles at $x_n^{\bar{x}}(t)$, with momenta $P_n^{\bar{x}}(t)$

at time t. The system has a momentum density

$$T^{\bar{\mu}\bar{0}}(x^{\bar{X}}) \equiv \sum_n P_n^{\bar{\mu}}(t)\delta(\vec{r} - \vec{r}_n(t))$$

$$= \sum_n P_n^{\bar{\mu}}(t)\frac{dx_n^{\bar{0}}(t)}{dt}\delta(\vec{r} - \vec{r}_n(t)). \qquad (7.20)$$

Its current is defined by

$$T^{\bar{\mu}\bar{i}}(x^{\bar{X}}) \equiv \sum_n P_n^{\bar{\mu}}(t)\frac{dx_n^{\bar{i}}(t)}{dt}\delta(\vec{r} - \vec{r}_n(t)), \quad \text{thus,}$$

$$T^{\bar{\mu}\bar{\nu}}(x^{\bar{X}}) \equiv \sum_n P_n^{\bar{\mu}}(t)\frac{dx_n^{\bar{\nu}}(t)}{dt}\delta(\vec{r} - \vec{r}_n(t)), \qquad (7.21)$$

$$P_n^{\bar{\nu}}(t) = m_n U_n^{\bar{\nu}}(t) = m_n\frac{dx_n^{\bar{\nu}}(t)}{d\tau}$$

$$= m_n(1 - (v_n)^2)^{-1/2}\frac{dx_n^{\bar{\nu}}(t)}{dt} = E_n\frac{dx_n^{\bar{\nu}}(t)}{dt},$$

$$T^{\bar{\mu}\bar{\nu}}(x^{\bar{X}}) = \sum_n \frac{P_n^{\bar{\mu}}(t)P_n^{\bar{\nu}}(t)}{E_n(t)}\delta(\vec{r} - \vec{r}_n(t)), \quad \text{so,}$$

$$T^{\bar{\mu}\bar{\nu}} = T^{\bar{\nu}\bar{\mu}}, \quad \text{symmetric.} \qquad (7.22)$$

The next proof shows that $T^{\bar{\mu}\bar{\nu}}$ is a tensor of rank 2. Write Eq. (7.21) as an integral over the four-dimensional (4D) delta function,

$$T^{\bar{\mu}\bar{\nu}}(x^{\bar{X}}) = \sum_n \int dt' P_n^{\bar{\mu}}(t')\frac{dx_n^{\bar{\nu}}(t')}{dt'}\delta(\vec{r} - \vec{r}_n(t'))\delta(t' - t)$$

$$= \sum_n \int d\tau P_n^{\bar{\mu}}(\tau)\frac{dx_n^{\bar{\nu}}(\tau)}{d\tau}\delta(\vec{r} - \vec{r}_n(\tau))\delta(\tau - t)$$

$$= \sum_n \int d\tau P_n^{\bar{\mu}}(\tau)U_n^{\bar{\nu}}(\tau)\delta(\vec{r} - \vec{r}_n(\tau))\delta(\tau - t).$$

On the right-hand side of the first line dt' cancels. Thus t' can be replaced by the invariant τ, when one changes variables. The right-hand side of the last equation is a contravariant tensor of rank 2 because on the right-hand side it is multiplied by the invariant $d\tau$ and by the invariant 4D delta function. As this is a tensor equation, $T^{\mu\nu}$ is a symmetric tensor of rank 2 in any frame.

Take a partial derivative of this tensor with respect to a spatial coordinate. From Eq. (7.21),

$$T^{\bar{\mu}\bar{i}}(x^{\bar{X}})_{,\bar{i}} = \sum_n P_n{}^{\bar{\mu}}(t)\frac{dx_n{}^{\bar{i}}(t)}{dt}\delta(\vec{r}-\vec{r}_n(t))_{,\bar{i}}$$

$$= \sum_n P_n{}^{\bar{\mu}}(t)\frac{dx_n{}^{\bar{i}}(t)}{dt}\frac{\partial}{\partial x^{\bar{i}}}\delta(\vec{r}-\vec{r}_n(t))$$

$$= -\sum_n P_n{}^{\bar{\mu}}(t)\frac{dx_n{}^{\bar{i}}(t)}{dt}\frac{\partial}{\partial x_n{}^{\bar{i}}(t)}\delta(\vec{r}-\vec{r}_n(t))$$

$$= -\sum_n P_n{}^{\bar{\mu}}(t)\delta(\vec{r}-\vec{r}_n(t))_{,t}, \quad \text{however,}$$

$$(f(t)g)_{,t} = \frac{df(t)}{dt}g + f(t)g_{,t}, \quad -f(t)g_{,t} = -(f(t)g)_{,t} + \frac{df(t)}{dt}g,$$

$$T^{\bar{\mu}\bar{i}}(x^{\bar{X}})_{,\bar{i}} = -T^{\bar{\mu}\bar{0}}(x^{\bar{X}})_{,\bar{0}} + \sum_n \frac{dP_n{}^{\bar{\mu}}(t)}{dt}\delta(\vec{r}-\vec{r}_n(t)),$$

$$T^{\bar{\mu}\bar{\nu}}(x^{\bar{X}})_{,\bar{\nu}} = \sum_n \frac{dP_n{}^{\bar{\mu}}(t)}{dt}\delta(\vec{r}-\vec{r}_n(t)).$$

Since $\frac{dP_n{}^{\bar{\mu}}(t)}{dt}$ is proportional to the force on the particle, the right-hand side of the above equation is just the density of the force. For free particles $T^{\bar{\mu}\bar{\nu}}{}_{,\bar{\nu}} = 0$. That is conservation of momentum. However, in this inertial frame,

$$T^{\bar{\mu}\bar{\nu}}{}_{,\bar{\nu}} = T^{\bar{\mu}\bar{\nu}}{}_{;\bar{\nu}} = 0 = T^{\mu\nu}{}_{;\nu}. \qquad (7.23)$$

The above is a tensor equation and holds in any frame. The energy–momentum tensor has the properties enumerated, when the Einstein equation Eq. (5.9) was discussed. If the particles were charged, they would interact via long range electric and magnetic forces. An additional term $T^{\mu\nu}_{E\&M}$ could be added to our tensor and the sum of these tensors would have zero divergence.

7.7 Quadrupole Radiation

An important case occurs when the source varies harmonically in time $S = 4T^{\mu\nu} = j_\omega[\vec{r}\,']\exp[-i\omega t']$. With the source near the origin, the observation point $\vec{r}$ may be in one of three zones: near, intermediate, and far. Each zone allows for different approximations. The far zone, where $d \ll \lambda \ll r$, is of interest to us. A violent event, triggering a gravitational wave, is likely to occur far from us. Here d is the source size, and is much smaller than the wavelength of the radiation. That, in turn, is much less than the radial coordinate of the observation point. Then,

$$|\vec{r} - \vec{r}\,'| = (r^2 + r'^2 - 2\vec{r} \cdot \vec{r}\,')^{1/2} = r(1 - 2\vec{r} \cdot \vec{r}\,'/r^2 + (r'/r)^2)^{1/2}$$

$$\approx r(1 - \vec{r} \cdot \vec{r}\,'/r^2) = r - \hat{e}_r \cdot \vec{r}\,',$$

$$\bar{h}^{\mu\nu}[\vec{r}, t] = \int_\infty dV' j_\omega[\vec{r}\,'] \frac{\exp[-i\omega(t - [r - \hat{e}_r \cdot \vec{r}\,'])]}{r - \hat{e}_r \cdot \vec{r}\,'}$$

$$\approx \frac{\exp[i(kr - \omega t)]}{r} \int_\infty dV' j_\omega[\vec{r}\,'] \exp[-ik\hat{e}_r \cdot \vec{r}\,']$$

$$\approx \frac{\exp[i(kr - \omega t)]}{r} \int_\infty dV' j_\omega[\vec{r}\,'], \text{ to lowest order}, \tag{7.24}$$

$$= \frac{4}{r} \int_\infty dV' T^{\mu\nu}[\vec{r}\,', t - r]. \tag{7.25}$$

The far zone is approximately locally inertial because it is so far from the source. In natural units $2\pi/\lambda = k = \omega$, but kr and ωt are written so that the equations look familiar. For the harmonic dependence, the solution Eq. (7.24) looks like an outgoing spherical wave with amplitude given by the integral. Recall that $g_{\mu\nu} = \eta_{\mu\nu} + h_{\mu\nu}$, thus the solution for $h^{\mu\nu}$ and $\bar{h}^{\mu\nu}$ are expressed in terms of the rectangular coordinates. Raising and lowering indices is done by $\eta^{\mu\nu}$ and $\eta_{\mu\nu}$. After these manipulations are carried out, the amplitudes can be expressed in other coordinate systems.

From energy conservation, the lowest order approximation yields

$$0 = T^{\mu\nu}{}_{;\nu} = T^{\mu\nu}{}_{,\nu} + \Gamma^\mu_{\xi\nu} T^{\xi\nu} + \Gamma^\nu_{\nu\xi} T^{\mu\xi} \approx T^{\mu\nu}{}_{,\nu}$$

$$= T^{0\nu}{}_{,\nu} = T^{00}{}_{,0} + T^{0k}{}_{,k}$$

$$= (T^{00}{}_{,0} + T^{0k}{}_{,k})_{,0} = T^{00}{}_{,0\,,0} + T^{0k}{}_{,k\,,0},$$

$$T^{00}{}_{,0\,,0} = (-T^{0k}{}_{,0})_{,k} = -(-T^{jk}{}_{,j})_{,k} = T^{jk}{}_{,j\,,k},$$

$$\int_\infty dV x^i x^n T^{00}{}_{,0\,,0} = \int_\infty dV x^i x^n T^{jk}{}_{,j\,,k}.$$

Integrate the integral on the right-hand side of the last equation by parts twice. The surface terms, far from the origin, vanish because $T^{jk} = 0$ there.

Thus,

$$\int_\infty dV\, x^i x^n T^{00},_{0,0} = \int_\infty dV\, x^i x^n [T^{jk},_j],_k$$

$$= -\int_\infty dV\, [T^{jk},_j][x^i x^n],_k$$

$$= -\int_\infty dV\, [T^{jk},_j][x^i,_k x^n + x^i x^n,_k]$$

$$= -\int_\infty dV\, [T^{jk},_j][\delta^i{}_k x^n + \delta^n{}_k x^i]$$

$$= \int_\infty dV\, T^{jk}[\delta^i{}_k x^n,_j + \delta^n{}_k x^i,_j]$$

$$= \int_\infty dV\, T^{jk}(\delta^i{}_k \delta^n{}_j + \delta^n{}_k \delta^i{}_j)$$

$$= \int_\infty dV\,(T^{ni} + T^{ni}) = 2\int_\infty dV\, T^{in}.$$

Using this result in Eq. (7.25) yields

$$r\bar{h}^{in}[\vec{r}, t]/2 = \int_\infty dV'\, x'^i x'^n (T^{00}[\vec{r}\,', t - r]),_{0,0}$$

$$= \frac{d^2 \int_\infty dV'\, x'^i x'^n T^{00}[\vec{r}\,', t - r]}{dt^2} \equiv \frac{d^2 I^{in}}{dt^2} = \frac{d^2 I_{in}}{dt^2}. \qquad (7.26)$$

Since the wave function is a real quantity, the *real part* of Eq. (7.26) is the solution. This leads to quadrupole radiation because $\bar{h}_{\mu\nu}$ is a tensor of rank 2. In electromagnetism, in lowest order, dipole radiation is possible because A_μ is a tensor of rank 1.

7.8 Gravity Wave Flux and Power

The wave flux is its energy/area/time. In natural units it is just m^{-2}. This quantity is integrated over the area of a sphere. That determines the power P or the wave luminosity L. As the source loses energy, it changes, and that observation can be used to detect the wave. The result from electromagnetic waves cannot be taken over directly as their amplitudes are tensors of rank 1, while a gravity wave amplitude is a tensor of rank 2. Thus, while the flux is still proportional to the absolute square of the amplitude, the all important proportionality factor is different. In order to calculate it, the approach of B. Schutz (2009) is followed.

Consider a plane transverse traceless wave moving in the z-direction. The flux transferred to an approximately continuous array of oscillators, elemental springs, is calculated. The springs are aligned along the x-direction in the plane $z = 0$. The springs have natural length l_0, equal masses m, small spring constant $m\omega_0^2/2$, and small damping constant $m\gamma$. The number of springs per unit area is $\frac{dn}{dA} = \alpha$. As the oscillators acquire energy, the wave loses energy, and its amplitude decreases. The relationship between flux and amplitude is found, not to depend on the springs, but is a property of the wave. The springs are just used as calculation facilitators.

Let the origin be at a spring's center with the masses at $x_{1,2}$. In flat space, the equations of motion for the masses are as follows:

$$\frac{d^2 x_2}{dt^2} = -\omega_0^2 (x_2 - x_1 - l_0)/2 - \gamma \frac{d(x_2 - x_1)}{dt},$$

$$\frac{d^2 x_1}{dt^2} = \omega_0^2 (x_2 - x_1 - l_0)/2 + \gamma \frac{d(x_2 - x_1)}{dt}, \qquad (7.27)$$

$$\frac{d^2 (x_2 - x_1 - l_0)}{dt^2} = -\omega_0^2 (x_2 - x_1 - l_0) - 2\gamma \frac{d(x_2 - x_1 - l_0)}{dt}.$$

One element of the wave $^{TT}\bar{h}_{xx}$ is considered. The other elements, and there must be other elements, since the trace is zero, would be handled in the same manner. When the wave is encountered, Eq. (7.17) yields the proper length between the masses,

$$l = (x_2 - x_1)(1 + {}^{TT}\bar{h}_{xx}/2)$$

$$\approx x_2 - x_1 + l_0 \, {}^{TT}\bar{h}_{xx}/2,$$

$$x_2 - x_1 = l - l_0 \, {}^{TT}\bar{h}_{xx}/2.$$

The terms on the right-hand side of Eq. (7.27) are unmodified by $^{TT}\bar{h}_{xx}$ because they are already expressed in terms of small quantities. This leads to

$$0 = \frac{d^2 ((l - l_0) - l_0 \, {}^{TT}\bar{h}_{xx}/2)}{dt^2} + \omega_0^2 (l - l_0) + 2\gamma \frac{d(l - l_0)}{dt}$$

$$\equiv \frac{d^2 \xi}{dt^2} + \omega_0^2 \xi + 2\gamma \frac{d\xi}{dt} - \frac{l_0}{2} \frac{d^2 \, {}^{TT}\bar{h}_{xx}}{dt^2}. \qquad (7.28)$$

This wave element can be expressed as

$$^{TT}\bar{h}_{xx} = \bar{A}\cos\Omega(t - z). \tag{7.29}$$

Insertion of Eq. (7.29) into Eq. (7.28) yields a differential equation with a sinusoidal solution,

$$0 = \frac{d^2\xi}{dt^2} + \omega_0^2\xi + 2\gamma\frac{d\xi}{dt} + (l_0\bar{A}\Omega^2/2)\cos\Omega(t - z),$$

$$\xi \equiv B\cos[\Omega(t - z) - \epsilon].$$

In order to obtain the constants B and ϵ, plug the solution into the differential equation. What multiplies both $\cos\Omega(t - z)$ and $\sin\Omega(t - z)$ must vanish,

$$0 = B(\omega_0^2 - \Omega^2)[\cos\Omega(t - z)\cos\epsilon + \sin\Omega(t - z)\sin\epsilon]$$

$$- 2B\gamma\Omega[\sin\Omega(t - z)\cos\epsilon - \cos\Omega(t - z)\sin\epsilon]$$

$$+ (l_0\bar{A}\Omega^2/2)\cos\Omega(t - z)$$

$$= B\sin\Omega(t - z)[\sin\epsilon(\omega_0^2 - \Omega^2) - 2\gamma\Omega\cos\epsilon]$$

$$= \cos\Omega(t - z)(B[(\omega_0^2 - \Omega^2)\cos\epsilon + 2\gamma\Omega\sin\epsilon] + l_0\bar{A}\Omega^2/2),$$

$$\tan\epsilon = \frac{2\gamma\Omega}{\omega_0^2 - \Omega^2}, \quad \sin\epsilon, \; \cos\epsilon = \frac{2\gamma\Omega, \; \omega_0^2 - \Omega^2}{[(2\gamma\Omega)^2 + (\omega_0^2 - \Omega^2)^2]^{1/2}}, \tag{7.30}$$

$$B = -(l_0\bar{A}\Omega^2/2)[(2\gamma\Omega)^2 + (\omega_0^2 - \Omega^2)^2]^{-1/2}, \; B/\bar{A} \ll 1. \tag{7.31}$$

Each oscillator is responding with a steady oscillation $B\cos(\Omega t - \epsilon)$. The energy dissipated by friction is compensated by the work done on the spring by the gravitational forces of the wave. Multiplying Eq. (7.28) by $md\xi/2$ yields

$$0 = \frac{m}{2}d\xi\left[\frac{d^2\xi}{dt^2} + \omega_0^2\xi\right] + m\gamma d\xi\frac{d\xi}{dt} - (ml_0/4)d\xi\frac{d^2\,^{TT}\bar{h}_{xx}}{dt^2}.$$

Each term in the equation has units of energy. The energy dissipated by friction is $dE = m\gamma d\xi\frac{d\xi}{dt}$. The rate of that energy change averaged over a period T is

$$\left\langle \frac{dE}{dt} \right\rangle = \frac{m\gamma}{T} \int_0^T dt \left(\frac{d\xi}{dt} \right)^2$$

$$= \frac{m\gamma\Omega^2 B^2}{T} \int_{-z}^{T-z} d(t-z) \sin^2(\Omega(t-z) - \epsilon)$$

$$= \frac{m\gamma\Omega^2 B^2}{2}.$$

As the wave passes through the $z = 0$ plane, all the oscillators take part. This yields the time averaged decrease in the flux,

$$\langle \delta F \rangle = -\frac{m\gamma\alpha\Omega^2 B^2}{2}. \tag{7.32}$$

It is small as $|B| \ll 1$.

The change in the wave amplitude is now required. The oscillations of the oscillator masses are also a source of a gravitational wave. The energy density is dominated by the mass term. For an oscillator, at a distance $r = (\rho^2 + z^2)^{1/2}$, from the observation point on the z-axis,

$$T_{00} = \sum_{i=1}^{2} m_i \delta[\vec{r}\,' - \vec{r}_i] = m([\delta(x' - x_1) + \delta(x' - x_2)]\delta(y' - y_0)\delta(z'))$$

$$= m([\delta(x' - [-l_0 - (l - l_0)]/2) + \delta(x' - [l_0 + (l - l_0)]/2)]\delta(y' - y_0)\delta(z'))$$

$$= m([\delta(x' - [-l_0 - B\cos(\Omega(t-r) - \epsilon)]/2)$$

$$+ \delta(x' - [+l_0 + B\cos(\Omega(t-r) - \epsilon)]/2)]\delta(y' - y_0)\delta(z')).$$

From Eq. (7.25), $\bar{h}^{00} = constant/r$ because the integral of a one-dimensional delta function is zero or one. This is not a wave that takes energy to very faraway points, and similarly for $\bar{h}^{22}$. Due to presence of $\delta(z')$, the elements $\bar{h}^{i3} = 0$. Due to the two delta functions $\delta(x' - x_0) + \delta(x' + x_0)$, the element $\bar{h}^{12} = 0$. So the only nonzero term is $\bar{h}^{11} = \bar{h}^{xx}$, where

$$\delta\bar{h}^{xx} = \frac{2}{r}\frac{d^2 I^{xx}}{dt^2} = \frac{m}{r}\frac{d^2[l_0 + B\cos(\Omega(t-r) - \epsilon)]^2}{dt^2}$$

$$\approx -\frac{2ml_0 B\Omega^2}{r}\cos(\Omega(t-r) - \epsilon). \tag{7.33}$$

In the last equation, the B^2 term is neglected.

At a point on the z-axis, the contribution from all of the oscillators is

$$\bar{h}_{xx} = -2Bl_0 m\Omega^2 \int_0^\infty \alpha 2\pi\rho d\rho \, \cos[\Omega(t-r) - \epsilon]/r,$$

$$\rho d\rho = (r^2 - z^2)^{1/2} d(r^2 - z^2)^{1/2} = r dr,$$

$$\bar{h}_{xx} = -4\pi\alpha Bl_0 m\Omega^2 \int_z^\infty \cos[\Omega(t-r) - \epsilon] dr$$

$$= 4\pi\alpha Bl_0 m\Omega \, \sin[\Omega(t-r) - \epsilon]|_z^\infty$$

$$= -4\pi\alpha Bl_0 m\Omega \, \sin[\Omega(t-z) - \epsilon].$$

In the last line, $\sin\infty = 0$ has been used.

Since the incident wave was in the TT gauge, the above must be put in that gauge. Then it can be combined with the original wave, to calculate the reduction in amplitude, when passing through the plane of the oscillators. The only nonzero spatial component of k_μ is $k_3 = \Omega$. Thus from Eqs. (7.10) and (7.11), using $B/\bar{A} \ll 1$, and that there is a single amplitude,

$$^{TT}\bar{h}_{xx} = \bar{h}_{xx} - \bar{h}_i{}^i/2 = \bar{h}_{xx}/2,$$

$$^{TT}\bar{h}_{xx}(\text{net}) = \bar{A}\cos(\Omega(t-z)) - 2\pi\alpha Bl_0 m\Omega \sin(\Omega(t-z) - \epsilon) \quad (7.34)$$

$$= \bar{A}[(1 + 2\pi\alpha(B/\bar{A})l_0 m\Omega \sin\epsilon)\cos(\Omega(t-z))$$

$$- 2\pi\alpha(B/\bar{A})l_0 m\Omega \sin(\Omega(t-z))\cos\epsilon].$$

Consider the following quantity:

$$f \equiv (\bar{A} + 2\pi\alpha Bl_0 m\Omega \sin\epsilon)[\cos(\Omega(t-z) + \Psi)]$$

$$= (\bar{A} + 2\pi\alpha Bl_0 m\Omega \sin\epsilon)[\cos(\Omega(t-z))\cos\Psi - \sin(\Omega(t-z))\sin\Psi]$$

$$= \bar{A}\cos\Psi(1 + 2\pi\alpha(B/\bar{A})l_0 m\Omega \sin\epsilon)[\cos(\Omega(t-z)) - \sin(\Omega(t-z))\tan\Psi].$$

The above is approximately the same as Eq. (7.34) because you can take,

$$\tan\Psi = \frac{(2\pi\alpha(B/\bar{A})l_0 m\Omega \cos\epsilon}{1 + 2\pi\alpha(B/\bar{A})l_0 m\Omega \sin\epsilon} \ll 1, \text{ thus, } \cos\Psi \approx 1,$$

$$^{TT}\bar{h}_{xx}(\text{net}) \approx (\bar{A} + 2\pi\alpha Bl_0 m\Omega \sin\epsilon)\cos(\Omega(t-z) - \Psi).$$

So apart from a small phase shift Ψ, the change in the wave amplitude is $\delta\bar{A} = 2\pi\alpha Bl_0 m\Omega \sin\epsilon$. Using Eqs. (7.30)–(7.32) gives

$$\frac{\delta\langle F\rangle}{\delta\bar{A}} = -\frac{m\gamma\alpha\Omega^2 B^2/2}{2\pi\alpha Bl_0 m\Omega \sin\epsilon} = -\frac{\gamma\Omega B}{4\pi l_0 \sin\epsilon}$$

$$= \frac{\gamma\Omega(l_0/2)\bar{A}\Omega^2/[(2\gamma\Omega)^2 + (\omega_0^2 - \Omega^2)^2]^{1/2}}{4\pi l_0(2\gamma\Omega)/[(2\gamma\Omega)^2 + (\omega_0^2 - \Omega^2)^2]^{1/2}} = \frac{1}{16\pi}\Omega^2\bar{A},$$

$$32\pi\langle F\rangle = \Omega^2\bar{A}^2 = \Omega^2\left\langle{}^{TT}\bar{h}_{xx}\;{}^{TT}\bar{h}^{xx}\right\rangle = \left\langle\frac{d^{TT}\bar{h}_{xx}}{dt}\frac{d^{TT}\bar{h}^{xx}}{dt}\right\rangle.$$

Though oscillators were used, the last equation has only wave properties; the oscillators have disappeared. They were just used as facilitators.

In general, a wave traveling in the z-direction, has more than just an $\bar{h}^{xx}$ amplitude, but all the amplitudes give similar results. From Eqs. (7.11) and (7.15),

$$ {}^{TT}\bar{h}_{xx} = -\,{}^{TT}\bar{h}_{yy} = \bar{h}_{xx} - \bar{h}/2 = \bar{h}_{xx} - (\bar{h}_{xx} + \bar{h}_{yy})/2 $$

$$ = \frac{1}{2}(\bar{h}_{xx} - \bar{h}_{yy}) = \frac{1}{r}\frac{d^2(I_{xx} - I_{yy})}{dt^2}, $$

$$ {}^{TT}\bar{h}_{xy} = {}^{TT}\bar{h}_{yx} = \bar{h}_{xy} = \bar{h}_{yx} = \frac{2}{r}\frac{d^2 I_{xy}}{dt^2}. $$

So, for a wave traveling in the z-direction, the general form for the flux is

$$\langle F\rangle = \frac{1}{32\pi}\left\langle\frac{d\,{}^{TT}\bar{h}_{ij}}{dt}\frac{d\,{}^{TT}\bar{h}^{ij}}{dt}\right\rangle. \tag{7.35}$$

Now define,

$$\bar{I}_{ij} \equiv I_{ij} - \delta_{ij}I_k{}^k/3. \tag{7.36}$$

One can prove that $\bar{I}$ is traceless and $I_{ij} = \bar{I}_{ij}$ if the amplitudes are in the transverse traceless gauge. The above equations can then be written,

$$ {}^{TT}\bar{h}_{xx} = -\,{}^{TT}\bar{h}_{yy} = \frac{1}{r}\frac{d^2(\bar{I}_{xx} - \bar{I}_{yy})}{dt^2}, $$

$$ {}^{TT}\bar{h}_{xy} = {}^{TT}\bar{h}_{yx} = \frac{2}{r}\frac{d^2\bar{I}_{xy}}{dt^2}, $$

$$32\pi\langle F\rangle = \left\langle \frac{d\,^{TT}\bar{h}_{xx}}{dt}\frac{d\,^{TT}\bar{h}^{xx}}{dt} + \frac{d\,^{TT}\bar{h}_{xy}}{dt}\frac{d\,^{TT}\bar{h}^{xy}}{dt}\right\rangle$$

$$+ \left\langle \frac{d\,^{TT}\bar{h}_{yx}}{dt}\frac{d\,^{TT}\bar{h}^{yx}}{dt} + \frac{d\,^{TT}\bar{h}_{yy}}{dt}\frac{d\,^{TT}\bar{h}^{yy}}{dt}\right\rangle,$$

$$16\pi r^2\langle F\rangle = \left\langle \frac{d^3(\bar{I}_{xx}-\bar{I}_{yy})}{dt^3}\frac{d^3(\bar{I}^{xx}-\bar{I}^{yy})}{dt^3} + 4\frac{d^3\bar{I}_{xy}}{dt^3}\frac{d^3\bar{I}^{xy}}{dt^3}\right\rangle.$$

$$(7.37)$$

It is easiest to calculate the radiated power with $\bar{I}_{ij}$. Also, it doesn't lead to radiation, if a spherically symmetric mass distribution oscillates radially. This is the Birkhoff theorem. It shows that in empty space, an object giving rise to a Schwarzschild metric, will still yield a static metric, if it oscillates radially. Problem 11 leads one through the proof, presented in Weinberg (1972).

Waves typically travel radially outward from the source, and not just in the z-direction. The luminosity is the normally outward flux, integrated over the surface of a sphere, of radial coordinate r, centered at the source. Equation (7.37) is not yet written in a form, that makes obvious, the outward wave flux at any point on the sphere. However, it can be put into such a form with a little manipulation. On the z-axis at the sphere's surface, the normal outward wave is in the z-direction. Therefore, one can rewrite Eq. (7.37) in the desired form. Use $0 = \bar{I}_{xx} + \bar{I}_{yy} + \bar{I}_{zz}$ and obtain

$$16\pi r^2\langle F\rangle = \left\langle 2\frac{d^3\bar{I}_{ij}}{dt^3}\frac{d^3\bar{I}^{ij}}{dt^3} - 4\frac{d^3\bar{I}_{zj}}{dt^3}\frac{d^3\bar{I}^{zj}}{dt^3} + \frac{d^3\bar{I}_{zz}}{dt^3}\frac{d^3\bar{I}^{zz}}{dt^3}\right\rangle. \quad (7.38)$$

The first product in Eq. (7.38) is independent of the integration point (θ,ϕ) on the surface. The products with the index z are particular to the integration point. They arose because at an arbitrary point, the outward normal unit vector $\hat{n}$ has components $n^i = n_i = x^i/r$. If the point is on the z-axis, then $n^{1,2,3} = (0,0,1)$ and $\frac{d^3\bar{I}_{jz}}{dt^3} = n^i\frac{d^3\bar{I}_{ji}}{dt^3}$. In general then, Eq. (7.37) can be written,

$$16\pi r^2\langle F\rangle = \left\langle 2\frac{d^3\bar{I}_{ij}}{dt^3}\frac{d^3\bar{I}^{ij}}{dt^3} - 4n^i n_k\frac{d^3\bar{I}_{ij}}{dt^3}\frac{d^3\bar{I}^{kj}}{dt^3}\right\rangle$$

$$+ \left\langle n^i n^j n_l n_m\frac{d^3\bar{I}_{ij}}{dt^3}\frac{d^3\bar{I}^{lm}}{dt^3}\right\rangle, \quad (7.39)$$

$$P = L = \int_{-1}^{1} d\cos\theta \int_{0}^{2\pi} d\phi\, r^2\langle F\rangle.$$

As an illustration, if the point was on the x-axis, the normally outward wave in the x-direction is required. Therefore, $n^{1,2,3} = (1,0,0)$, and $n^i \frac{d^3 \bar{I}_{ij}}{dt^3} = \frac{d^3 \bar{I}_{xj}}{dt^3}$,

$$16\pi r^2 \langle F \rangle = \left\langle 2\frac{d^3 \bar{I}_{ij}}{dt^3}\frac{d^3 \bar{I}^{ij}}{dt^3} - 4\frac{d^3 \bar{I}_{xj}}{dt^3}\frac{d^3 \bar{I}^{xj}}{dt^3} + \frac{d^3 \bar{I}_{xx}}{dt^3}\frac{d^3 \bar{I}^{xx}}{dt^3} \right\rangle.$$

As expected, the above has the same form as Eq. (7.38) with x replacing z.

When integrating over the solid angle, only the unit vectors need to be integrated. The $\bar{I}$ terms are independent of the integration point. The integrals are easily calculated,

$$n^{1,2,3} = (\sin\theta\cos\phi, \sin\theta\sin\phi, \cos\theta),$$

$$\int_{-1}^{1} d\cos\theta \int_{0}^{2\pi} d\phi = 4\pi,$$

$$\int_{-1}^{1} d\cos\theta \int_{0}^{2\pi} d\phi\, n^i n_k = \frac{4\pi}{3}\delta^i{}_k,$$

$$\int_{-1}^{1} d\cos\theta \int_{0}^{2\pi} d\phi\, n^i n^j n_l n_m = \frac{4\pi}{15}(\delta^{ij}\delta_{lm} + \delta^i{}_m\delta^j{}_l + \delta^i{}_l\delta^j{}_m).$$

Using these results the luminosity is

$$4L = \left\langle 2\frac{d^3 \bar{I}_{ij}}{dt^3}\frac{d^3 \bar{I}^{ij}}{dt^3} - \frac{4}{3}\frac{d^3 \bar{I}_{ij}}{dt^3}\frac{d^3 \bar{I}^{ij}}{dt^3} \right\rangle$$

$$+ \left\langle \frac{1}{15}\left(\frac{d^3 \bar{I}^i{}_i}{dt^3}\frac{d^3 \bar{I}^l{}_l}{dt^3} + 2\frac{d^3 \bar{I}_{ij}}{dt^3}\frac{d^3 \bar{I}^{ij}}{dt^3}\right) \right\rangle, \tag{7.40}$$

$$L = \frac{1}{4}\left(2 - \frac{4}{3} + \frac{2}{15}\right)\left\langle \frac{d^3 \bar{I}_{ij}}{dt^3}\frac{\bar{I}^{ij}}{dt^3} \right\rangle = \frac{1}{5}\left\langle \frac{d^3 \bar{I}_{ij}}{dt^3}\frac{d^3 \bar{I}^{ij}}{dt^3} \right\rangle.$$

7.9   Binary Neutron Star System Radiation

The binary pulsar PSR B1913+16, shown schematically in Fig. 7.3, was discovered by R. Hulse and J. H. Taylor (Hulse, 1975). This occurred in 1973 at the Arecibo radio telescope. It is in a gravitational bound state with an unseen neutron star. The pulsar is a magnetized neutron star, whose rapid rotation generates a plasma, the source of beamed radio waves. They are seen at earth, as periodic pulses, every 0.059 s. This is because the radio waves are beamed along the magnetic axis, but that axis rotates about the spin axis of the star. The rotation period of such a massive compact body

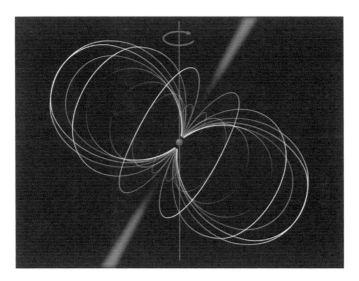

Fig. 7.3 The magnetic axis of a pulsar, along which radio waves are beamed, is rotating about the pulsar rotation axis. The beam is periodically in the line of sight of an earth observer.

is very stable against external perturbations. It is actually a very accurate clock. Modern timing devices can measure the period with high precision. Search the WWW for "pulsar" and you'll find some wonderful images. Neutron stars are highly compact massive objects. They are supported against gravitational collapse by neutron degeneracy, a purely quantum effect. For a mass of $1.4M_s$, the neutron star surface radial coordinate is ≈ 10–20 km.

Upon discovery, it was noted that the pulsing rate varied. This was interpreted as due to the pulsar traveling in a changing gravitational field, as it orbited an unseen neighbor. The pulsar was tracked for decades, and the parameters of the orbit were obtained from the slight changes in the pulsing rate. The orbiting is a source of a gravitational wave. The wave carries away energy, reflected in changes in the orbit. Since the pulsar is a radio emitter, the experimenters have to remove the distortion, due to the index of refraction of the intergalactic medium. An optical pulsar would have been simpler.

The discoverers were joined by J. M. Weisberg who performed much of the data analysis. Their paper Weisberg (2010) and references therein, describe the intricacies of extracting the orbit parameters. They found that this is a wonderful system with which to test GR. For example, the advance of the periastron is ≈35,000 times that of the perihelion of Mercury. The periastron is the distance of closest approach to its unseen neighbor. However, the prize here is the detection of a gravitational wave carrying energy away from the system.

The pulsar orbital period was measured at periastron over the course of decades. The best-fit parameters extracted from the data are: masses $m_1 \approx m_2 \approx 1.4 M_s$; distance from earth $r = 6400$ parsec (pc), where 1 pc = 3.3 ly $= 3.1 \times 10^{16}$ m; period $T = 7.75$ h; eccentricity $e = 0.617$; and semi-major axis $a = 1.95 \times 10^9$ m. The prize was obtained through the detection of a decrease in the period $\frac{dT}{dt} = -(2.4056 \pm 0.0051) \times 10^{-12}$. The excellent agreement with the GR calculation is shown in Fig. 7.4. This "indirect" confirmation of gravity waves led to the 1993 Nobel Prize for the discoverers.

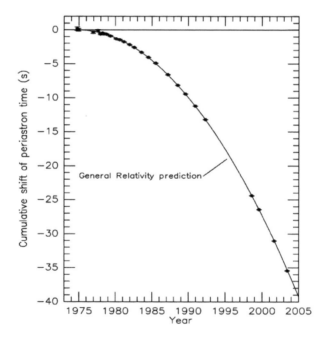

Fig. 7.4   The cumulative shift of the time of periastron, the closest approach distance between the neutron stars, of the Hulse–Taylor pulsar PSR B1913+16 binary system. The data points agree with the solid curve GR calculation to within 0.33%.

The cumulative shift in the time of periastron is a particularly illustrative way to express the data. Weisberg used for this quantity (see Suzuki, 2019)

$$\Delta T_P = \frac{dT}{dt}\frac{1}{2T}T_P^2.$$

Here, T_P is the time between the start and the last periastron, and ΔT_P is the cumulative shift. Try using $T_P = 30$ y and see that $\Delta T_P \approx -40$ s.

Some texts, that call attention to this system, treat the orbits as circular. This gives a result off by an order of magnitude from the correct result because the eccentricity is large. Though more calculation is required, the Newtonian elliptical orbit calculation will be used in this text. It was first worked out by P. C. Peters (1964), although in a different notation, and with many steps left to the reader.

Before beginning the calculation one should review the properties of the two-body Newtonian system (see Problem 6). The gravitational wave amplitude is expected to be very small, so Newtonian orbits adequately specify the star positions. The reduced mass μ travels in an elliptical orbit in the $\theta = \pi/2$ plane, with focus at the center of mass, taken at rest. Its position $\vec{r} = \rho[\phi]\hat{e}_\rho$, period T, and energy of the system E, are obtained from,

$$m \equiv m_1 + m_2, \quad \mu \equiv m_1 m_2/m,$$

$$\rho^2\frac{d\phi}{dt} = \frac{J}{\mu} = [ma(1-e^2)]^{1/2}, \tag{7.41}$$

$$\rho = [a(1-e^2)]/(1 + e\cos\phi), \tag{7.42}$$

$$T = \frac{2\pi a^{3/2}}{m^{1/2}}, \quad a = \left(\frac{m^{1/2}T}{2\pi}\right)^{2/3}, \tag{7.43}$$

$$E = -\frac{1}{2}\frac{\mu m}{a} = -\frac{\mu}{2}\left(\frac{2\pi m}{T}\right)^{2/3}, \tag{7.44}$$

$$\frac{dT}{dt} = \frac{3\pi}{m^{1/2}}a^{1/2}\frac{da}{dt} = \frac{3}{(2\pi)^{2/3}}\frac{T^{5/3}}{\mu m^{2/3}}\frac{dE}{dt}, \tag{7.45}$$

$$\frac{dE}{dt} = \frac{1}{2}\frac{\mu m}{a^2}\frac{da}{dt}, \tag{7.46}$$

where J is a constant angular momentum, a is the semi-major axis, and e is the eccentricity. In Newton's theory of the two-body system $\frac{da}{dt} = 0$, but in GR the gravitational waves, generated by the motion of the masses, take power out of the system. The energy decreases, becoming more negative, with time. So a and T decrease with time. Such changes can be compared with observation to test Einstein's theory.

The wave power calculation needs the energy–momentum tensor, assumed dominated by the mass density. The latter is used to get the integrals I_{ij},

$$T_{00} = \sum_{i=1}^{2} m_i \delta \left(x' - \vec{r}_{ix}^{\,R} \right) \delta \left(y' - \vec{r}_{iy}^{\,R} \right) \delta \left(z' \right),$$

$$(7.47)$$

$$I_{xy,xx,yy} = \sum_{i=1}^{2} m_i (\vec{r}_{ix}^{\,R} \vec{r}_{iy}^{\,R}, \ (\vec{r}_{ix}^{\,R})^2, \ (\vec{r}_{iy}^{\,R})^2).$$

The superscript R means the retarded value, that when the wave left the binary. From now on, all the variables a, e, ϕ have retarded values, and the superscript is dropped. The mass positions are obtained from that of the reduced mass,

$$\vec{r}_{i(x,y)} = (-1)^i \frac{\mu}{m_i} \vec{r}_{(x,y)} = (-1)^i \frac{\mu}{m_i} \rho(\cos\phi, \sin\phi). \qquad (7.48)$$

The second derivatives, with respect to time of the I's, are needed for the wave amplitude. The third derivatives are used to calculate the luminosity. The time derivatives of a and e are due to the gravitational wave radiation. They are $\ll \frac{d\phi}{dt}$. Thus, the latter is the only time derivative that counts, and it is given by Eq. (7.41).

From Eqs. (7.41) and (7.42), I_{xx} is

$$I_{xx} = \sum_{i=1}^{2} m_i \left(\frac{\mu}{m_i} \right)^2 [a(1-e^2)]^2 \left(\frac{\cos\phi}{1 + e\cos\phi} \right)^2$$

$$= \mu^2 \left(\sum_{i=1}^{2} \frac{1}{m_i} \right) [a(1-e^2)]^2 \left(\frac{\cos\phi}{1 + e\cos\phi} \right)^2$$

$$= \mu \, [a(1-e^2)]^2 \left(\frac{\cos\phi}{1 + e\cos\phi} \right)^2.$$

The derivatives are as follows:

$$\frac{dI_{xx}}{dt} = \mu \left[a \left(1 - e^2 \right) \right]^2 \left(\frac{-2 \cos \phi \sin \phi}{(1 + e \cos \phi)^2} + \frac{2e \cos^2 \phi \sin \phi}{(1 + e \cos \phi)^3} \right) \frac{d\phi}{dt}$$

$$= -\mu \left[ma \left(1 - e^2 \right) \right]^{1/2} \frac{\sin 2\phi}{1 + e \cos \phi},$$

$$\frac{d^2 I_{xx}}{dt^2} = \frac{2\mu m}{a \left(1 - e^2 \right)} \left[\sin^2 \phi - \cos^2 \phi - e \cos^3 \phi \right], \tag{7.49}$$

$$\frac{d^3 I_{xx}}{dt^3} = \frac{2\mu m^{3/2}}{\left[a \left(1 - e^2 \right) \right]^{5/2}} \left(1 + e \cos \phi \right)^2$$

$$\times \left[2 \sin 2\phi + 3e \sin \phi \cos^2 \phi \right]. \tag{7.50}$$

The yy terms are as follows:

$$I_{yy} = \mu \left[a \left(1 - e^2 \right) \right]^2 \left(\frac{\sin \phi}{1 + e \cos \phi} \right)^2 = \frac{\mu \left[a \left(1 - e^2 \right) \right]^2}{(1 + e \cos \phi)^2} - I_{xx},$$

$$\frac{dI_{yy}}{dt} = 2\mu \left[ma \left(1 - e^2 \right) \right]^{1/2} \frac{e \sin \phi}{1 + e \cos \phi} - \frac{dI_{xx}}{dt},$$

$$\frac{d^2 I_{yy}}{dt^2} = \frac{2\mu m}{a \left(1 - e^2 \right)} \left(e^2 + e \cos \phi + \cos 2\phi + e \cos^3 \phi \right), \tag{7.51}$$

$$\frac{d^3 I_{yy}}{dt^3} = \frac{-2\mu m^{3/2}}{\left[a \left(1 - e^2 \right) \right]^{5/2}} \left(1 + e \cos \phi \right)^2$$

$$\times \left[2 \sin 2\phi + e \sin \phi \left(1 + 3 \cos^2 \phi \right) \right]. \tag{7.52}$$

The xy terms are as follows:

$$I_{xy} = \mu \left[a \left(1 - e^2 \right) \right]^2 \frac{\sin \phi \cos \phi}{(1 + e \cos \phi)^2} = \mu \left[a \left(1 - e^2 \right) \right]^2 \frac{\sin 2\phi}{2 \left(1 + e \cos \phi \right)^2},$$

$$\frac{dI_{xy}}{dt} = \mu \left[ma \left(1 - e^2 \right) \right]^{1/2} \frac{\cos 2\phi + e \cos \phi}{1 + e \cos \phi},$$

$$\frac{d^2 I_{xy}}{dt^2} = -\frac{\mu m}{a \left(1 - e^2 \right)} \sin \phi \left[4 \cos \phi + e \left(1 + \sin^2 \phi + 3 \cos^2 \phi \right) \right], \tag{7.53}$$

$$\frac{d^3 I_{xy}}{dt^3} = \frac{-2\mu m^{3/2}}{\left[a \left(1 - e^2 \right) \right]^{5/2}} \left(1 + e \cos \phi \right)^2$$

$$\times \left[2 \cos 2\phi - e \cos \phi \left(1 - 3 \cos^2 \phi \right) \right]. \tag{7.54}$$

The wave equation solutions are as follows:

$$\bar{h}_{ij} = \frac{2}{r}\frac{d^2 I_{ij}}{dt^2} = \frac{4\mu m}{a\left(1-e^2\right)r}f_{ij}\left[\phi\right], \qquad (7.55)$$

where f_{xx}, f_{yy}, f_{xy} are functions of ϕ and are given in Eqs. (7.49), (7.51), and (7.53). Using the best-fit orbit parameters, the factor multiplying f_{ij} is $\approx 0.7 \times 10^{-22}$. This shows where the small amplitudes expected by LIGO come from.

The luminosity can be calculated using Eqs. (7.36) and (7.40),

$$\bar{I}_{ij} \equiv I_{ij} - \frac{1}{3}\delta_{ij}I_k{}^k = I_{ij} - \frac{1}{3}\delta_{ij}\left(I_{xx}+I_{yy}\right),$$

$$\bar{I}_{xx} = I_{xx} - \frac{1}{3}\left(I_{xx}+I_{yy}\right) = \frac{1}{3}\left(2I_{xx}-I_{yy}\right),$$

$$\bar{I}_{yy} = I_{yy} - \frac{1}{3}\left(I_{xx}+I_{yy}\right) = -\frac{1}{3}\left(I_{xx}-2I_{yy}\right),$$

$$\bar{I}_{zz} = -\frac{1}{3}\left(I_{xx}+I_{yy}\right),$$

$$\bar{I}_{xy} = I_{xy}.$$

The luminosity becomes

$$L = \frac{1}{5}\left\langle \frac{d^3\bar{I}_{ij}}{dt^3}\frac{d^3\bar{I}^{ij}}{dt^3}\right\rangle,$$

$$9\frac{d^3\bar{I}_{ij}}{dt^3}\frac{d^3\bar{I}^{ij}}{dt^3} = 18\frac{d^3 I_{xy}}{dt^3}\frac{d^3 I^{xy}}{dt^3}$$

$$+ \left(2\frac{d^3 I_{xx}}{dt^3}-\frac{d^3 I_{yy}}{dt^3}\right)\left(2\frac{d^3 I^{xx}}{dt^3}-\frac{d^3 I^{yy}}{dt^3}\right)$$

$$+ \left(2\frac{d^3 I_{yy}}{dt^3}-\frac{d^3 I_{xx}}{dt^3}\right)\left(2\frac{d^3 I^{yy}}{dt^3}-\frac{d^3 I^{xx}}{dt^3}\right)$$

$$+ \left(\frac{d^3 I_{xx}}{dt^3}+\frac{d^3 I_{yy}}{dt^3}\right)\left(\frac{d^3 I^{xx}}{dt^3}+\frac{d^3 I^{yy}}{dt^3}\right),$$

$$L = \frac{8\mu^2 m^3}{15\left[a\left(1-e^2\right)\right]^5}$$

$$\times \left\langle (1+e\cos\phi)^4(e^2\sin^2\phi+12[1+e\cos\phi]^2)\right\rangle. \qquad (7.56)$$

In order to average over a period, use Eq. (7.41), to change the integration variable from t to ϕ,

$$\langle L \rangle = \frac{8\mu^2 m^3}{15\left[a\left(1-e^2\right)\right]^5}$$

$$\times \frac{1}{T} \int_0^T dt (1 + e\cos\phi)^4 (e^2 \sin^2\phi + 12[1 + e\cos\phi]^2),$$

$$dt = \frac{\rho^2}{\left[ma\left(1-e^2\right)\right]^{1/2}} d\phi,$$

$$\langle L \rangle = \frac{8\mu^2 m^3}{15\left[a\left(1-e^2\right)\right]^5} \frac{m^{1/2}}{a^{3/2}} \frac{\left[a\left(1-e^2\right)\right]^2}{\left[ma\left(1-e^2\right)\right]^{1/2}}$$

$$\times \frac{1}{2\pi} \int_0^{2\pi} (1 + e\cos\phi)^2 \left[e^2 \sin^2\phi + 12\left(1 + e\cos\phi\right)^2\right] d\phi$$

$$= \frac{4\mu^2 m^3}{15\pi a^5 \left(1-e^2\right)^{7/2}}$$

$$\times \int_0^{2\pi} d\phi \left(1 + e\cos\phi\right)^2 \left[12 + 24e\cos\phi + e^2 \left(1 + 11\cos^2\phi\right)\right].$$

However,

$$\int_0^{2\pi} d\phi \cos^{0,1,2,3,4}\phi = 2\pi, 0, \pi, 0, \frac{3}{4}\pi.$$

So the final answer is

$$\langle L \rangle = \frac{32}{5} \frac{\mu^2 m^3}{a^5} \frac{1 + \frac{73}{24}e^2 + \frac{37}{96}e^4}{\left(1-e^2\right)^{7/2}} \equiv \frac{32}{5} \frac{\mu^2 m^3}{a^5} F[e]$$

$$= \frac{32}{5} \left(\frac{2\pi}{T}\right)^{10/3} \mu^2 m^{4/3} F[e] = \frac{2}{5} \left(\frac{2\pi m}{T}\right)^{10/3} F[e]. \qquad (7.57)$$

Since $e = 0.617$, the enhancement factor due to the eccentricity is $F[e] = 11.84$. Thus, the luminosity and the period change are a factor 11.84 times greater in magnitude than would be calculated for the circular orbit case.

The change in the period, obtained using Eqs. (7.45) and (7.57), yields

$$-\frac{dE}{dt} = \langle L \rangle,$$

$$\frac{dT}{dt} = -\frac{3}{(2\pi)^{2/3}} \frac{T^{5/3}}{\mu m^{2/3}} \frac{32}{5} (2\pi)^{10/3} \frac{\mu^2 m^{4/3}}{T^{10/3}} F[e]$$

$$= -3 (2\pi)^{8/3} \frac{32}{5} \frac{\mu m^{2/3}}{T^{5/3}} F[e] = -\frac{48\pi}{5} \left(\frac{2\pi m}{T}\right)^{5/3} F[e] \qquad (7.58)$$

$$= -30.16 \times 11.84 \left(\frac{2\pi 2 \times 1.4 \times 1.484 \times 10^3}{7.75 \times 3.6 \times 10^3 \times 3 \times 10^8}\right)^{5/3}$$

$$= -2.4 \times 10^{-12} = 76 \ \mu s/y.$$

This result agrees with the data within 0.33%. The above expression can be used to find T and a, after any elapsed time, since $\frac{dT}{dt} \propto -T^{-5/3}$.

It should be noted that these gravitational waves cannot be seen by LIGO. At the frequencies that are of order T^{-1}, the LIGO noise is much larger than 10^{-22}. When the two stars get close to merging, the signal will be huge because they are so close. However, as Problem 8 indicates, this is hundreds of million years in the future. There was an announcement in October, 2017 of LIGO spotting such a merger in August, 2017. Unlike the merger of two black holes that results in a final black hole, which can only be detected via gravitational waves, this merger was also seen by optical, radio, and gamma-ray burst telescopes. The merger event was 1.3 Mly away, or about 60 times farther away, than the system here considered. The merger produced a kilonova. The line spectra from heavy elements like gold were detected. This confirmed the model of heavy element creation, expected from such explosions.

Problems

1. Start with Eq. (7.1) and show the steps that lead to Eq. (7.2). Starting with Eq. (7.3) show the steps leading to Eq. (7.4). Show that the second term in Eq. (7.5) is equal to $-\Box h_{\beta\nu}$.
2. In showing that GR admits a wave equation solution, the utility of a gauge transformation and use of the trace inverse was noted,

$$h_{\mu'\nu'} = h_{\mu\nu} - \xi_{\mu,\nu} - \xi_{\nu,\mu},$$

$$\bar{h}_{\mu\nu} = h_{\mu\nu} - \eta_{\mu\nu}h/2 = h_{\mu\nu} - \eta_{\mu\nu}\eta^{\chi\beta}h_{\beta\chi}/2.$$

(a) Show $\bar{h}^{\mu\nu} = \bar{h}^{\mu'\nu'} + \eta^{\chi\nu}\xi^{\mu}{}_{,\chi} + \eta^{\chi\mu}\xi^{\nu}{}_{,\chi} - \eta^{\mu\nu}\xi^{\alpha}{}_{,\alpha}$.

(b) Show that if $\bar{h}^{\mu\nu}{}_{;\nu} = \bar{h}^{\mu\nu}{}_{,\nu} \neq 0$, the above gauge transform makes $\bar{h}^{\mu'\nu'}{}_{;\nu'} = 0$, with the correct gauge vector choice. What is the choice?

3. Prove $\delta(t' - t) = \frac{1}{2\pi}\int_{-\infty}^{\infty} d\omega \exp[i\omega(t' - t)]$.

4. Prove that if $\bar{I}_{ij} \equiv I_{ij} - \delta_{ij}I_k{}^k/3$, then $\text{Trace}(\bar{I}) = \sum_i \bar{I}_{ii} = \bar{I}_i{}^i = 0$. Then prove that for a wave traveling in the z-direction, Eq. (7.38) and Eq. (7.37) are the same.

5. Prove

$$\int_{-1}^{1} d\cos\theta \int_0^{2\pi} d\phi\, n^i n_k = 4\pi\delta^i{}_k/3, \quad \text{and,}$$

$$\int_{-1}^{1} d\cos\theta \int_0^{2\pi} d\phi\, n^i n^j n_l n_m = 4\pi(\delta^{ij}\delta_{lm} + \delta^i{}_m\delta^j{}_l + \delta^i{}_l\delta^j{}_m)/15.$$

6. Obtain Eqs. (7.41)–(7.46) for the orbit of the Hulse–Taylor pulsar from Newtonian physics.

7. Show for the Hulse–Taylor pulsar that the wavelength of the gravitational wave is much larger than the source size and much smaller than the distance to the observation point. This justifies use of quadrupole radiation for the power in the far zone.

8. Calculate the period as a function of time for the Hulse–Taylor pulsar. How many years will it take for the semi-major axis a to be reduced to half of the observed value and for a to fall to one-tenth of its observed value?

9. Suppose the stars of the Hulse–Taylor binary system traveled in the same circular orbit of radius R such that they are at opposite points of the diameter. In this case, calculate the amplitudes $\bar{h}_{ij}$ and show that the frequency of the gravitational wave is twice that of the star rotation.

10. Use the results from Problem 9 and carry out the calculation of the power output. Show that you obtain the last form of Eq. (7.57) with $F[e] = 1$.

11. To prove the Birkhoff theorem, consider a spherically symmetric, time varying gravitational field in empty space. It has a metric in the form of Eq. (5.11) with the functions A, B, C, D now functions of r and t. Show that a transform to new coordinates,

$$r' = r[D(r,t)]^{1/2}, \ dt' \text{ is of the form } F(r',t)dt - G(r',t)dr',$$

where $F(r',t) = t',_{t}$, $-G(r',t) = t',_{r'}$, and $F,_{r'} = -G,_{t'}$ are such as to get rid of the $dr'dt'$ term. The metric, after dropping the primes, is then, given by Eq. (5.15), except now, $\Phi = \Phi(r,t)$ and $\Delta = \Delta(r,t)$. Find the nonzero C symbols that, due to the time dependence, did not appear in Eq. (5.16). Next, show that $R_{03} \propto \Delta,_{t}$ and so in free space, $\Delta = \Delta(r)$, and $g_{33} = 1/(1 - 2M'/r)$. Next calculate the terms in R_{00} and R_{33} that have time derivatives. Show they are directly proportional to ($\Delta,_{t}$ or $\Delta,_{t}$ $,_{t}$) and so vanish in free space. Thus, as in the stationary case, $0 = R_{00} = R_{33}$ yields $(g_{00}g_{33}),_{r} = 0$, or, $g_{00} = -f(t)(1 - 2M'/r)$. Finally, show that if a new time, $dt'' = dt[f(t)]^{1/2}$, is defined, the Schwarzschild metric is obtained. Therefore, a spherically symmetric pulsating gravitational field does not produce a gravitational wave that can carry energy infinitely faraway.

Chapter 8

Black Holes and Kerr Space

8.1 Static Black Holes

In the era before SR and GR, there was speculation about possible compact spherically symmetric objects of large mass M' and radius R from which even light could not escape. The incorrect argument was made on the basis of the escape speed v_E, of an object of mass M, using Newtonian mechanics. The escape speed condition is that the total energy vanishes at $r = R$. Then the object stops at $r = \infty$. This yields

$$M(v_E)^2/2 - MM'/R = 0, \quad v_E = (2M'/R)^{1/2}.$$

So when $R = 2M'$, $v_E = 1$, and light would be bound to the compact object.

The connection to GR is easy to see. It is just where the Schwarzschild metric has a singularity other than $r = 0$. As spherical coordinates will be used, let $r^p \equiv (r)^p$. The metric is

$$(d\tau)^2 = (1 - 2M'/r)(dt)^2 - (1 - 2M'/r)^{-1}(dr)^2$$
$$- r^2[(d\theta)^2 + \sin^2\theta(d\phi)^2]$$
$$= (1 - R/r)(dt)^2 - (1 - R/r)^{-1}(dr)^2 - r^2[(d\theta)^2 + \sin^2\theta(d\phi)^2].$$

Thus,

$$1 - R/r = 0, \quad (1 - R/r)^{-1} = \infty \quad \text{when } r = R,$$

where R is called the Schwarzschild radius. For an object with the mass of the sun, it has a very small value $R = 2M_s = 2.968 \times 10^3$ m. In the case of the sun, such a radius is well within the sun's radius, and wouldn't contain much of the sun's mass. A black hole, however, is a real singularity at $r = 0$, and R is external to it. In the region accessible to observation $R/r < 1$, the applications of the Schwarzschild metric found in Chapter 6 apply. However, to emphasize that black holes are spoken of, $2M'$ will be replaced by R for the rest of this chapter.

The apparent singularity at R is not real and is due to the choice of coordinates. This can be seen by recalling that in deriving the Schwarzschild metric, the Ricci tensor $R_{\mu\nu}$ vanished in vacuum. Thus, there can't be a real singularity at the vacuum point $r = R$. One can seek other coordinates that make the apparent singularity disappear. The following ones, known as the Kruskal coordinates (Kruskal, 1960) (r', t'), do the trick:

$$r'^2 - t'^2 = K^2(r/R - 1)\exp[r/R], \tag{8.1}$$

$$2r't'/(r'^2 + t'^2) = \tanh[t/R], \ t = R\tanh^{-1}[2r't'/(r'^2 + t'^2)]. \tag{8.2}$$

For given r, Eq. (8.1) yields hyperbolas in the (r', t') plane, as illustrated in Fig. 8.1. If $r > R$, the hyperbolas are symmetric about the r'-axis. They cross that axis at positive values of r', that increase as $r \to \infty$. If $r < R$, they are symmetric about the t'-axis. They cross that axis at positive values of t', that increase as $r \to 0$. The singularity at $r = 0$ in these coordinates is a hyperbola. However, since it is a singularity, it isn't completely understood by current physics.

When $r = R$, $t' = \pm r'$. These are lines with slopes ± 1 that define the asymptotic behavior of the hyperbolas. For given t, Eq. (8.2) allows introduction of constant K',

$$K' = \tanh[t/R], \quad 0 \le |K'| \le 1, \tag{8.3}$$

$$0 = K'r'^2 - 2r't' + K't'^2,$$

$$t' = (2r' \pm [(2r')^2 - (2K'r')^2]^{1/2})/(2K')$$

$$= (r'/K')[1 \pm (1 - K'^2)^{1/2}]. \tag{8.4}$$

These lines start at the origin. They extend in the positive r'-direction with slope $[1\pm(1-K'^2)^{1/2}]/K'$. Note that when $t = \pm\infty$, $K' = \pm 1$. They overlap the lines of $r = R$. A massive particle, shown schematically by the dashed curve in Fig. 8.1, travels from $r \ge R$ to $r = R$. It will cut through all the time lines, and get there at $t = \infty$. This time is measured on the clock

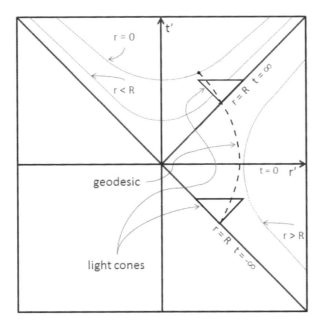

Fig. 8.1 Kruskal coordinates (r', t') for a black hole. In these coordinates, the only singularity is at $r = 0$. The light cones are erect and as wide open as in SR. Once light or any massive particle gets to $r < R$, it can never get to $r > R$ again.

of a faraway at-rest observer. However, the time on a clock moving with the particle, the particle's proper time, proceeds as if nothing unusual was happening. As the light cones in these coordinates are, as erect and open as in SR, once the object gets to $r < R$ it must go to the singularity, and similarly for photons.

In order to see this, the metric must be written in terms of r', t'. A little manipulation of Eq. (8.2) yields,

$$dt = R\frac{(r'^2 + t'^2)2(t'dr' + r'dt') - 2r't'[2(r'dr' + t'dt')]}{[1 - (2r't'/[r'^2 + t'^2])^2](r'^2 + t'^2)^2}$$

$$= 2R\frac{dr't'((r'^2 + t'^2 - 2r'^2) + dt'r'(r'^2 + t'^2 - 2t'^2)}{(r'^2 + t'^2)^2 - (2r't')^2}$$

$$= 2R(r'^2 - t'^2)^{-1}[-t'dr' + r'dt'],$$

$$(dt)^2 = (2R)^2\frac{t'^2(dr')^2 + r'^2(dt')^2 - 2r't'dr'dt'}{(r'^2 - t'^2)^2}$$

$$= (2R)^2 \frac{t'^2(dr')^2 + r'^2(dt')^2 - 2r't'dr'dt'}{K^4(r/R - 1)^2 \exp[2r/R]}$$

$$= 4R^4 \frac{t'^2(dr')^2 + r'^2(dt')^2 - 2r't'dr'dt'}{K^4 r^2 (1 - R/r)^2 \exp[2r/R]},$$

$$(1 - R/r)(dt)^2 = 4R^4 \frac{t'^2(dr')^2 + r'^2(dt')^2 - 2r't'dr'dt'}{K^4 r^2 (1 - R/r) \exp[2r/R]}. \tag{8.5}$$

Similarly manipulation of Eq. (8.1) yields

$$2(r'dr' - t'dt') = (K^2/R) \exp[r/R](1 + r/R - 1)dr$$

$$= (K/R)^2 r \exp[r/R]dr,$$

$$(dr)^2 = 4R^4 \frac{(r'dr')^2 + (t'dt')^2 - 2r't'dr'dt'}{K^4 r^2 \exp[2r/R]},$$

$$\frac{(dr)^2}{1 - R/r} = 4R^4 \frac{(r'dr')^2 + (t'dt')^2 - 2r't'dr'dt'}{K^4 r^2 (1 - R/r) \exp[2r/R]}. \tag{8.6}$$

Combining Eqs. (8.5) and (8.6) gives

$$(1 - R/r)(dt)^2 - \frac{(dr)^2}{1 - R/r} = 4R^4 \frac{[r'^2 - t'^2][(dt')^2 - (dr')^2]}{K^4 r^2 (1 - R/r) \exp[2r/R]}$$

$$= \frac{4R^3[(dt')^2 - (dr')^2]}{K^2 r \exp[r/R]}.$$

This yields the relation for the element of proper time,

$$d\tau^2 = \frac{4R^3[(dt')^2 - (dr')^2]}{K^2 r \exp[r/R]} - r^2[(d\theta)^2 + \sin^2\theta(d\phi)^2]. \tag{8.7}$$

In Eq. (8.7), r is not regarded as a coordinate, but as a function of r' and t'. One can see that there is no singularity at $r = R$. The metric is well defined as long as r^2 is positive definite. There is a singularity at $r = 0$, as expected.

The condition for light signals is $d\tau = 0$. Thus, for radial travel $d\theta = d\phi = 0$, one obtains $dt' = \pm dr'$. This means that the light cones shown in Fig. 8.1 are vertically erect and fully open, as in SR. Massive object world lines would move inside the cones because their speed is less than that of light. So once $r < R$, massive objects and light move towards increasing t', inside and on the surfaces of the cones. They must get to the singularity. The surface at R is a one-way membrane called an event horizon. As the object moves towards $r = R$, it may send out light signals to a faraway observer. The light may possess a given frequency, so that the crests are emitted at regular intervals of the object's proper time. The proper time is running slower and slower, compared to the time on a faraway clock. The time periods between intervals on the faraway clock become longer and longer. The light is red shifted to the extreme. Such signals disappear from the view of the faraway observer before the object reaches the horizon.

In recent years, observation has confirmed the existence of black holes. Stars near the center of our galaxy have been observed for decades. Their orbits indicate rotation about an unseen mass. Other sightings include the accretion of visible matter into a dark area. Such is often accompanied by the emission of X-rays and gamma rays as the gravitational pulls on the matter are so violent. Gravitational waves indicating the mergers of two black holes have been detected, and are discussed in Chapter 7.

And at last, an actual picture of a black hole has been constructed. The Event Horizon Telescope Collaboration (2019) used a collection of very long baseline radio telescopes. They looked at the giant black holes in the center of our galaxy and galaxy M87. The latter is 55×10^6 ly away. Putting all the data together produced sufficient triangulation accuracy to construct the image shown in Fig. 8.2. The characteristics of this black hole were worked out by modeling extremely curved space, strong magnetic fields, and superheated matter. The mass was $(6.5 \pm 0.7) \times 10^9 M_s$ and the horizon radial coordinate was $\approx 2 \times 10^{13}$ m. The latter is about 2.5 times smaller

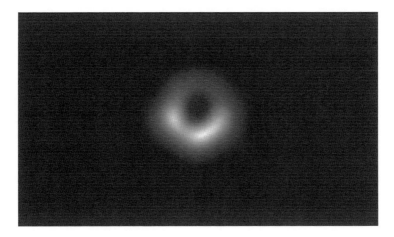

Fig. 8.2 First image of a black hole in M87 by the Event Horizon Telescope Collaboration.

than the size of the central dark region. The light surrounding this region is due to bound superheated matter and the bending of light. Since the light intensity isn't the same all around, the black hole must be rotating (see below).

Black holes arise from the supernova of massive stars. If the remnant mass is greater than the mass that could be supported by neutron degeneracy pressure, a purely quantum effect, nothing can prevent a complete collapse to a singularity. All that is left is an event horizon. This is such a special object that quantum questions are raised.

Consider the singularity. At first it seems to be a point of infinite density. We have other experience with infinite density. For example, electrons and muons have no structure, but from their motion in electromagnetic fields, have finite mass and charge. However, K. Thorne (2014) explained that there is no matter in a black hole. There is only the infinite warping of spacetime. The in-falling matter is crunched out of existence, leaving an event horizon. Our present physics just cannot handle this singularity. A quantum theory of gravity is required.

Black holes are predicted to have a finite life span. They evaporate with the emission of Hawking radiation [Hawking (1975)]. The radiation has a very low temperature black-body spectrum. This is a quantum effect in the region of extreme gravity, a still unsolved problem in general. Here, the process is very briefly described. At the event horizon, quantum fluctuations in the vacuum produce particle–anti-particle pairs. Normally, these particles would immediately recombine, before energy and momentum non-conservation could be observed. However, if one particle is just inside the horizon, it can't get out, while the other particle may be outside the horizon. It can escape from the black hole. The net effect to an outside observer, could such low temperatures be detected, is that a particle has emerged from, and mass has been lost by the black hole. So R has decreased. Eventually, the black hole will evaporate. Not to worry, the evaporation time of a solar mass black hole is orders of magnitude greater than the age of the universe.

8.2   Black Holes with Angular Momentum

R. P. Kerr found the metric for the space outside a spinning sphere. This occurred almost 50 years after Einstein published GR theory, so one gets a feel for the difficulty of the problem. The paper's length is a single page (Kerr, 1967). No doubt an elegant proof for a mathematician. However, it is not transparent for mere mortals like this author. Nor have I found a proof that is sufficiently transparent and brief to justify presentation in this text. Lengthy proofs are available, see Adler (1965) and Chandrasekhar (1983). For our purposes, I'll ask you to accept that the metric written in terms of spherical coordinates has the following form:

$$(d\tau)^2 = (1 - Rr/\Sigma)(dt)^2 + (2Rra\sin^2\theta/\Sigma)dtd\phi - (\Sigma/\Lambda)(dr)^2$$
$$- \Sigma(d\theta)^2 - \sin^2\theta(r^2 + a^2 + Rra^2\sin^2\theta/\Sigma)(d\phi)^2, \quad (8.8)$$

$$\Lambda = r^2 + a^2 - Rr, \quad \Sigma = r^2 + a^2\cos^2\theta. \quad (8.9)$$

When $a = 0$, the metric becomes that of a Schwarzschild black hole. One observes that the angular momentum forces an additional constant a into the metric. That constant has units of meters in our naturalized units, but that is also the units of angular momentum per unit mass, and it is interpreted as such.

Since $(d\tau)^2 = -g_{\mu\nu}dx^\mu dx^\nu$, the element of area $[g_{11}g_{22}]^{1/2}d\theta d\phi$ does not have the form of the element of area for a sphere $r^2\sin\theta d\theta d\phi$. This is another surprising feature of spacetime near a rotating black hole. The metric $g_{\mu\nu}$ has off-diagonal elements, but it is easy to find $g^{\mu\nu}$ using $g^{\xi\mu}g_{\xi\nu} = \delta^\mu_\nu$. Problem 5 asks for the proof of the following results:

$$g^{ii} = 1/g_{ii}, \quad i = 1, 3, \quad (8.10)$$

$$g^{02} = g^{20} = g_{02}/\Psi, \; g^{00} = -g_{22}/\Psi, \; g^{22} = -g_{00}/\Psi, \quad (8.11)$$

$$\Psi = \Lambda\sin^2\theta. \quad (8.12)$$

The natures of the singularities for a black hole with angular momentum are more complicated than those of a static black hole. There is a true singularity at $\Sigma = 0$, $(r, \theta) = (0, \pi/2)$, as every term in $(d\tau)^2$ blows up at that point except for the $(dr)^2$ and $(d\theta)^2$ terms, both of which vanish. You could also form the scalar $R^{\xi\mu\nu\chi}R_{\xi\mu\nu\chi}$ to see that it has the above singularity condition.

In the case of the static black hole, there was a coordinate singularity because the $(dr)^2$ term blew up at $r = R$. In this case, $g_{33} = \Sigma/\Lambda \to \infty$, when $\Lambda \to 0$,

$$0 = \Lambda = r^2 + a^2 - Rr = (r - R_+)(r - R_-),$$
$$r = R_\pm = R/2 \pm ([R/2]^2 - a^2)^{1/2}. \quad (8.13)$$

One observes that the maximum limit for a is $R/2$. This is called the static limit, and here $R_+ = R_-$. The inner horizon, given by R_-, is also called the Cauchy horizon. Both horizons are one way, you can cross only going in. The outside or event horizon is the boundary between the black hole and the outside world. If $a = 0$, it reduces to R, but in this case, it is smaller.

For a Schwarzschild black hole, $g_{00} = 0$ when $g_{33} = \infty$. However, for a rotating black hole, the surface where $g_{00} = 0$ occurs when

$$0 = \Sigma - Rr = r^2 - Rr + a^2 \cos^2 \theta = (r - r_+)(r - r_-),$$
$$r = r_\pm = R/2 \pm ([R/2]^2 - a^2 \cos^2 \theta)^{1/2}. \tag{8.14}$$

The values of $r_\pm$ depend on θ. In general $r_- \leq R_-$ and $r_+ \geq R_+$. The equality is at $|\cos \theta| = 1$. The region between R_+ and r_+ is called the ergosphere. In this region, all objects experience frame-drag (see Section 8.3). Due to the $\cos^2 \theta$ term, the outer extremity of the ergosphere is not a true sphere. It is flattened and equal to the event horizon at $|\cos \theta| = 1$, but otherwise extends beyond the event horizon, and is accessible to observation.

8.3 Frame-drag

8.3.1 *Simple Examples*

The metric elements are independent of x^0 and x^2. Thus, P_0 and P_2 are constants of the motion for massive particles. So even if the particle starts with $P_2 = 0$, but $P_0 \neq 0$, subsequently the angular velocity is nonzero,

$$P^0 = m\frac{dt}{d\tau} = g^{0\nu}P_\nu = g^{00}P_0,$$

$$P^2 = m\frac{d\phi}{d\tau} = g^{2\nu}P_\nu = g^{20}P_0,$$

$$\frac{d\phi}{dt} = \frac{d\phi}{d\tau} \bigg/ \frac{dt}{d\tau} = \frac{g^{20}}{g^{00}} = -\frac{g_{20}}{g_{22}}$$

$$= \frac{Rra}{\Sigma[r^2 + a^2] + Rra^2 \sin^2 \theta} > 0.$$

The particle will develop angular momentum and angular velocity because of the metric. The angular velocity is in the direction of the spinning mass. The spinning mass drags spacetime, so this effect is called frame-drag.

An interesting effect for light is predicted when $dr = d\theta = 0$, $\theta = \pi/2$, and $r = R$. For light $(d\tau)^2 = 0$,

$$0 = (1 - R/r)(dt)^2 + (2Ra/r)dtd\phi - (r^2 + a^2 + Ra^2/r)(d\phi)^2$$

$$= 1 - R/r + (2Ra/r)\frac{d\phi}{dt} - (r^2 + a^2 + Ra^2/r)\left(\frac{d\phi}{dt}\right)^2,$$

$$\frac{d\phi}{dt} = \frac{-2Ra/r \pm [(2Ra/r)^2 + 4(1 - R/r)(r^2 + a^2 + Ra^2/r)]^{1/2}}{-2[r^2 + a^2 + Ra^2/r]}$$

$$= \frac{a \mp a}{(R^2 + 2a^2)}.$$

The solution $\frac{d\phi}{dt} = 2a/(R^2 + 2a^2)$ represents light going in the same direction as the rotating black hole. As dt is the proper time of a faraway observer, the angular velocity according to that observer is constrained by the properties of the black hole. The other solution is zero, a rather unexpected result, if light is emitted in the direction opposite that of the black hole's rotation. A massive particle, that moves slower than light, would soon find itself rotating in the same direction as the black hole. This would occur even if it started rotating with arbitrarily large kinetic energy in the opposite direction.

However, one must realize that the faraway observer can neither confirm nor falsify this prediction. That observer cannot observe these at-rest photons, even in principle. If an effect can neither be confirmed nor falsified, then we are not speaking of a scientific prediction. As noted in Problem 2.15, the measurement of light speed requires a local observer, one armed with a proper-time clock and proper distance measuring device. That observer would experience severe frame-drag, move in the same direction as the black hole, and would observe the photon traveling with $c = \frac{dL_p}{d\tau} = 1$.

8.3.2 *An Experimental Test*

Gneral Relativity predicts that a gyroscope orbiting a non-spinning spherically symmetric mass will precess (geodetic precession) in its plane of motion. Frame-drag causes gyroscope precession perpendicular to the plane of motion, the Lense–Thirring effect (Lense and Thirring, 1918). L. Schiff's (1960) paper stimulated the Gravity Probe B (GP-B) experiment. This experiment was designed to test the Lense–Thirring prediction. The experimental predictions were calculated using Schiff's equation (3), but steps leading to it were not provided in Schiff's paper. GP-B was started about

50 years ago by NASA. The experiment (Everitt, 2011) is a classic of modern technology.

Four fused silica rotors served as the gyroscopes. They were machined to be spherical and homogeneous to one part per million. They were coated with superconducting niobium and kept at a temperature of 1.8 K(elvin). In orbit, the rotors were suspended electrically, and spun up with helium gas. The spinning niobium spheres develop a magnetic moment, and sensitive magnetometers mounted inside the housing could keep track of the spin direction. For calibration, there was also a star seeking telescope inside the housing. The device was put in a circular polar orbit about earth, where it fell freely for about a year. However, gravity is very weak near the earth's surface. The sought for precession is very small. Another five years were required to model the systematic errors, for example, the small non-spherical rotor contribution, in order to be confident of the result.

The experimental results, in arcseconds/year ($''$/y) are: geodetic precession, 6.602 ± 0.018 and frame-drag precession 0.0372 ± 0.0072. These are to be compared with the GR predictions: 6.606 and 0.0392. The geodetic result is in excellent agreement with GR. Frame-drag precession is a much smaller effect, and the large error is due to consideration of numerous sources of systematic error. So while in agreement with GR, it can't rule out competing theories.

8.4 Calculation of the GP-B Predictions

8.4.1 *Expansion of the Metric Elements*

In this section, the calculation of the predictions for the GP-B precessions will be carried out. It is too difficult to work with the full metric. One can expand the metric elements to higher than the lowest order correction explored in Chapter 5, and obtain meaningful results. Such calculations are the essence of what is called post-Newtonian celestial mechanics. In making these calculations, I have been guided, in part, by S. Weinberg (1972).

In Chapter 5, the weak gravity lowest order correction to the metric, for a static spherically symmetric source of gravitation, gave the Newtonian potential Ψ_G,

$$g_{00} = \eta_{00} + h_{00} = -1 - 2\Psi_G = -1 + 2M'/r, \qquad (8.15)$$

$$g_{ii} = 1 + h_{ii} = 1 - 2\Psi_G, \qquad (8.16)$$

$$\bar{v} \approx (M'/r)^{1/2} \ll 1. \qquad (8.17)$$

The g_{ii} value was obtained in Problem 5.2. The typical small speed was found to be $\bar{v}$. Here earth's mass $M_e = M'$ is the source mass, so that $R \to 2M_e$ in the Kerr metric. The above metric elements are of order $\bar{v}^2$ and because there will be higher order corrections, they are now written as $^2h_{00}$, $^2h_{ii}$.

In order to see how to expand the metric elements in powers of small $\bar{v}$, consider the Kerr metric, e.g.,

$$g_{00} = -(1 - 2M_e r/\Sigma) = \eta_{00} + 2M_e r/\Sigma = -1 + 2M_e r/(r^2 + a^2 \cos^2 \theta)$$

$$= -1 + (2M_e/r)(1 + [a/r]^2 \cos^2 \theta)^{-1}$$

$$= -1 + (2M_e/r)(1 - [a/r]^2 \cos^2 \theta + \cdots)$$

$$\equiv -1 + {}^2h_{00} + {}^4h_{00} + \cdots. \tag{8.18}$$

The superscript 4 at the left of h_{00} indicates the next higher order correction directly proportional to the $\bar{v}^4$. It is due to a being an angular momentum per unit mass. Thus a/r, is dimensionless in the natural system of units. An expansion in even powers of $\bar{v}$ is expected, as there is no sign change in g_{00}, when the sign of t changes.

Next consider the expansion of g_{ij}. In the Kerr metric, only $g_{ii} \neq 0$, e.g.,

$$g_{33} = \Sigma/\Lambda = (r^2 + a^2 \cos^2 \theta)(r^2 - 2M_e r + a^2)^{-1}$$

$$= (1 + [a/r]^2 \cos^2 \theta)(1 - 2M_e/r + [a/r]^2)^{-1}$$

$$= (1 + [a/r]^2 \cos^2 \theta)(1 + 2M_e/r - [a/r]^2 + [2M_e/r - [a/r]^2]^2 + \cdots)$$

$$= 1 + \frac{2M_e}{r}\left(1 - \frac{[a/r]^2}{[2M_e/r]} \sin^2 \theta\right)$$

$$+ \left(\frac{2M_e}{r}\right)^2 \times \left(1 - \frac{[a/r]^2}{[2M_e/r]}(2 - \cos^2 \theta) + \frac{[a/r]^4}{[2M_e/r]^2}\right) + \cdots.$$

Thus, the general expansion is

$$g_{ij} = \eta_{ij} + {}^2h_{ij} + {}^4h_{ij} + \cdots. \tag{8.19}$$

This is again an expansion in even powers of $\bar{v}$, for the same reason as for g_{00}.

For a black hole, $[a/r]^2/[R/r] < 1$ since $a/R \leq 1/2$. The earth's moment of inertia (query any search engine) is $\approx 0.829I$ (uniform sphere) $= 0.829(\frac{2}{5}M_e R_e^2)$, where R_e and ω_e are earth's radius and angular velocity,

$$[a/R_e]^2 = [I\omega_e/M_e]^2/R_e^2 = (0.829[2/5]R_e\omega_e)^2 = 0.27 \times 10^{-12},$$

$$2M_e/R_e = 1.4 \times 10^{-9}, \quad \text{thus } [a/R_e]^2/[2M_e/R_e] \approx 0.19 \times 10^{-3}.$$

The mass term dominates.

Finally, consider the metric element g_{0i}. In the Kerr metric, only g_{02} is nonzero,

$$g_{02} = -2M_e ra\sin^2\theta/r^2/(1 + [a/r]^2\cos^2\theta)$$

$$= -2[M_e/r][a/r]r\sin^2\theta(1 - [a/r]^2\cos^2\theta + [a/r]^4\cos^4\theta + \cdots).$$

So the general expansion is expected to be

$$g_{0i} = {}^3h_{0i} + {}^5h_{0i} + \cdots. \tag{8.20}$$

Here an expansion in odd powers of $\bar{v}$ is expected, as g_{0i} changes sign when t changes sign.

One sees that the lowest order term in the metric element, responsible for frame-drag, is directly proportional to the $\bar{v}^3$. This is sufficient for the GP-B experiment, and so the only terms kept satisfy,

$$\text{Term} \propto \bar{v}^n, \quad n \leq 3. \tag{8.21}$$

8.4.2   *The C Symbols*

The C symbols involve $g^{\mu\nu}$. They are also obtained as expansions in $\bar{v}^n$, that satisfy Eq. (8.21). From, $g^{\mu\xi}g_{\nu\xi} = \delta^\mu_\nu$,

$$1 = g^{0\xi}g_{0\xi} = g^{00}g_{00} + g^{0i}g_{0i} = (\eta^{00} + {}^2h^{00})(\eta_{00} + {}^2h_{00})$$

$$= 1 - {}^2h^{00} - {}^2h_{00}, \tag{8.22}$$

$${}^2h^{00} = - {}^2h_{00},$$

$$0 = g^{0\xi}g_{i\xi} = g^{00}g_{i0} + g^{0j}g_{ij} = {}^3h_{i0}\eta^{00} + {}^3h^{0j}\eta_{ij}, \tag{8.23}$$

$${}^3h^{i0} = {}^3h_{i0},$$

$$\delta^i_{\ j} = g^{i\xi}g_{j\xi} = g^{i0}g_{j0} + g^{ik}g_{jk} = (\eta^{ik} + {}^2h^{ik})(\eta_{jk} + {}^2h_{jk})$$

$$= \eta^{ik}\eta_{jk} + {}^2h_{jk}\eta^{ik} + {}^2h^{ik}\eta_{jk} = \delta^i_{\ j} + {}^2h_{jk}\eta^{ik} + {}^2h^{ik}\eta_{jk}, \tag{8.24}$$

$${}^2h^{ij} = - {}^2h_{ji}.$$

The C symbols involve terms with a partial derivative with respect to time. It's important to note that as for powers of $\bar{v}$,

$$\frac{\partial}{\partial t} \propto \bar{v}/r.$$

This result and Eqs. (8.18)–(8.20) are required to make sure Eq. (8.21) is satisfied. The C symbols are obtained from

$$\Gamma^\xi_{\mu\nu} = g^{\xi\chi}(g_{\mu\chi,\nu} + g_{\nu\chi,\mu} - g_{\mu\nu,\chi})/2.$$

Those with contravariant index 0 are as follows:

$$\Gamma^0_{00} = g^{0\chi}(g_{0\chi,0} + g_{0\chi,0} - g_{00,\chi})/2$$
$$= g^{00}(g_{00,0})/2 + g^{0i}(2g_{0i,0} - g_{00,i})/2, \qquad (8.25)$$
$${}^3\Gamma^0_{00} = \eta^{00}\,{}^2h_{00,0}/2 = -\,{}^2h_{00,0}/2,$$

$$\Gamma^0_{0i} = g^{0\chi}(g_{i\chi,0} + g_{0\chi,i} - g_{i0,\chi})/2$$
$$= g^{00}(g_{i0,0} + g_{00,i} - g_{i0,0})/2 + g^{0j}(g_{ij,0} + g_{0j,i} - g_{0i,j})/2, \quad (8.26)$$
$${}^2\Gamma^0_{0i} = \eta^{00}({}^2h_{00,i})/2 = -\,{}^2h_{00,i}/2,$$

$$\Gamma^0_{ij} = g^{0\chi}(g_{i\chi,j} + g_{j\chi,i} - g_{ij,\chi})/2$$
$$= g^{00}(g_{i0,j} + g_{j0,i} - g_{ij,0})/2 + g^{0k}(g_{ik,j} + g_{jk,i} - g_{ij,k})/2, \qquad (8.27)$$
$${}^3\Gamma^0_{ij} = -({}^3h_{i0,j} + {}^3h_{j0,i} - {}^2h_{ij,0})/2.$$

Those with contravariant index i are as follows:

$$\Gamma^i_{00} = g^{i\chi}(g_{0\chi,0} + g_{0\chi,0} - g_{00,\chi})/2$$
$$= g^{i0}g_{00,0}/2 + g^{ik}(g_{0k,0} + g_{0k,0} - g_{00,k})/2, \qquad (8.28)$$
$${}^2\Gamma^i_{00} = \eta^{ik}(-\,{}^2h_{00,k})/2 = -\,{}^2h_{00,i}/2,$$

$$\Gamma^i_{0j} = g^{i\chi}(g_{0\chi,j} + g_{j\chi,0} - g_{0j,\chi})/2$$
$$= g^{i0}(g_{00,j} + g_{j0,0} - g_{0j,0})/2 + g^{ik}(g_{0k,j} + g_{jk,0} - g_{0j,k})/2, \quad (8.29)$$
$${}^3\Gamma^i_{0j} = \eta^{ik}({}^3h_{0k,j} + {}^2h_{jk,0} - {}^3h_{0j,k})/2$$
$$= ({}^3h_{0i,j} + {}^2h_{ji,0} - {}^3h_{0j,i})/2,$$

$$\Gamma^i_{jk} = g^{i\chi}(g_{j\chi,k} + g_{k\chi,j} - g_{jk,\chi})/2$$
$$= g^{i0}(g_{j0,k} + g_{k0,j} - g_{jk,0})/2 + g^{im}(g_{jm,k} + g_{km,j} - g_{jk,m})/2, \qquad (8.30)$$
$${}^2\Gamma^i_{jk} = \eta^{im}({}^2h_{jm,k} + {}^2h_{km,j} - {}^2h_{jk,m})/2$$
$$= ({}^2h_{ji,k} + {}^2h_{ki,j} - {}^2h_{jk,i})/2.$$

8.4.3   *Ricci Tensor and Einstein Field Equations*

In order to express the various expansion terms of a metric, as potentials in terms of the energy momentum tensor, just like $^2h_{00} = -2\Psi_G$, the Einstein field equations are needed,

$$G_{\mu\xi} = R_{\mu\xi} - g_{\mu\xi}R/2 = 8\pi T_{\mu\xi},$$

$$g^{\mu\xi}(R_{\mu\xi} - g_{\mu\xi}R/2) = 8\pi g^{\mu\xi}T_{\mu\xi},$$

$$R - 2R = -R = 8\pi g^{\mu\xi}T_{\mu\xi}, \tag{8.31}$$

$$R_{\mu\nu} = 8\pi(T_{\mu\nu} - g_{\mu\nu}g^{\chi\xi}T_{\chi\xi}/2).$$

The nonzero Ricci tensor elements $R_{\mu\nu}$ will be evaluated so that Eq. (8.21) is obeyed. The reader should note that $R_{\mu\nu}$ in this text has the opposite sign of $R_{\mu\nu}$ in Weinberg's text. In some of the equations below, η^{ii} is used to remind the reader to sum over i.

$$R_{\mu\nu} = R^{\xi}{}_{\mu\xi\nu} = \Gamma^{\chi}_{\mu\nu}\Gamma^{\xi}_{\xi\chi} - \Gamma^{\chi}_{\xi\mu}\Gamma^{\xi}_{\nu\chi} + \Gamma^{\xi}_{\mu\nu},_{\xi} - \Gamma^{\xi}_{\xi\mu},_{\nu}.$$

However, the product $\Gamma\Gamma \propto \bar{v}^{>3}$ and may be neglected. Moreover,

$$R_{00} = \Gamma^{\xi}_{00},_{\xi} - \Gamma^{\xi}_{\xi 0},_0 = \Gamma^{i}_{00},_{i} - \Gamma^{i}_{i0},_0,$$

$$^2R_{00} = {}^2\Gamma^{i}_{00},_{i} = -\eta^{ii}\,{}^2h_{00},_{i}\,,_{i}/2 = -\nabla^2\,{}^2h_{00}/2. \tag{8.32}$$

$$R_{0i} = \Gamma^{\xi}_{0i},_{\xi} - \Gamma^{\xi}_{\xi 0},_{i} = \Gamma^{0}_{0i},_0 - \Gamma^{0}_{00},_{i} + \Gamma^{j}_{0i},_{j} - \Gamma^{j}_{j0},_{i} = \Gamma^{j}_{0i},_{j} - \Gamma^{j}_{j0},_{i},$$

$$^3R_{0i} = \eta^{jj}[(^3h_{0j},_{i} + {}^2h_{ij},_0 - {}^3h_{0i},_{j}),_{j} - (^3h_{0j},_{j} + {}^2h_{jj},_0 - {}^3h_{0j},_{j}),_{i}]/2$$

$$= \eta^{jj}(^2h_{ij},_0\,,_{j} - {}^3h_{0i},_{j}\,,_{j} - {}^2h_{jj},_0\,,_{i} + {}^3h_{0j},_{j}\,,_{i})/2$$

$$= [-\nabla^2\,{}^3h_{0i} + \eta^{jj}(^2h_{ij},_0\,,_{j} - {}^2h_{jj},_0\,,_{i} + {}^3h_{0j},_{j}\,,_{i})]/2. \tag{8.33}$$

$$R_{ij} = \Gamma^{0}_{ij},_0 - \Gamma^{0}_{i0},_{j} + \Gamma^{k}_{ij},_{k} - \Gamma^{k}_{ik},_{j},$$

$$^2R_{ij} = + {}^2h_{00},_{i}\,,_{j}/2 + \eta^{kk}[(^2h_{ki},_{j} + {}^2h_{jk},_{i} - {}^2h_{ij},_{k}),_{k}]/2$$

$$\qquad - \eta^{kk}[(^2h_{kk},_{i} + {}^2h_{ik},_{k} - {}^2h_{ki},_{k}),_{j}]/2$$

$$= {}^2h_{00},_{i}\,,_{j}/2 + \eta^{kk}[(^2h_{ki},_{j} + {}^2h_{jk},_{i} - {}^2h_{ij},_{k}),_{k}]/2$$

$$\qquad - \eta^{kk}[(^2h_{kk},_{i} + {}^2h_{ik},_{k} - {}^2h_{ki},_{k}),_{j}]/2$$

$$= [-\nabla^2\,{}^2h_{ij} + {}^2h_{00},_{i}\,,_{j}$$

$$\qquad + \eta^{kk}(^2h_{ki},_{j}\,,_{k} + {}^2h_{jk},_{i}\,,_{k} - {}^2h_{kk},_{i}\,,_{j})]/2. \tag{8.34}$$

Solution of the field equations with this form of Ricci tensor is quite difficult, but use of the result in Problem 3.4, simplifies things. There it was seen that one can always transform to a set of harmonic coordinates, that satisfy $g^{\mu\nu}\Gamma^\xi_{\mu\nu} = 0$. This equation must hold for every order in $\bar{v}$. Thus,

$$g^{\mu\nu}\Gamma^0_{\mu\nu} = g^{00}\Gamma^0_{00} + g^{i0}\Gamma^0_{i0} + g^{0j}\Gamma^0_{0j} + g^{ij}\Gamma^0_{ij},$$

$$0 = \eta^{00}\,{}^3\Gamma^0_{00} + \eta^{ij}\,{}^3\Gamma^0_{ij}$$

$$= -\eta^{00}\,{}^2h_{00,0}/2 - \eta^{ij}({}^3h_{i0,j} + {}^3h_{j0,i} - {}^2h_{ij,0})/2$$

$$= {}^2h_{00,0}/2 + \eta^{ii}(-{}^3h_{i0,i} + {}^2h_{ii,0}/2), \tag{8.35}$$

$$g^{\mu\nu}\Gamma^i_{\mu\nu} = g^{00}\Gamma^i_{00} + g^{k0}\Gamma^i_{k0} + g^{0j}\Gamma^i_{0j} + g^{kj}\Gamma^i_{kj},$$

$$0 = \eta^{00}\,{}^2\Gamma^i_{00} + \eta^{kj}\,{}^2\Gamma^i_{kj}$$

$$= -\eta^{00}\,{}^2h_{00,i}/2 + \eta^{kj}({}^2h_{ji,k} + {}^2h_{ki,j} - {}^2h_{jk,i})/2$$

$$= {}^2h_{00,i}/2 + \eta^{jj}({}^2h_{ji,j} - {}^2h_{jj,i}/2). \tag{8.36}$$

Differentiate Eq. (8.35) with respect to x^j and obtain

$$0 = {}^2h_{00,0\,,j}/2 + \eta^{ii}(-{}^3h_{i0,i\,,j} + {}^2h_{ii,0\,,j}/2),$$

$${}^2h_{00,0\,,j}/2 = -\eta^{ii}(-{}^3h_{i0,i\,,j} + {}^2h_{ii,0\,,j}/2). \tag{8.37}$$

Differentiate Eq. (8.36) with respect to x^0 and use Eq. (8.37), and obtain

$$0 = {}^2h_{00,i\,,0}/2 + \eta^{jj}({}^2h_{ji,j\,,0} - {}^2h_{jj,i\,,0}/2), \text{ use } i \leftrightarrow j,$$

$$= {}^2h_{00,j\,,0}/2 + \eta^{ii}({}^2h_{ij,i\,,0} - {}^2h_{ii,j\,,0}/2), \text{ use Eq. (8.37)},$$

$$= -\eta^{ii}(-{}^3h_{i0,i\,,j} + {}^2h_{ii,0\,,j}/2 - {}^2h_{ij,i\,,0} + {}^2h_{ii,j\,,0}/2)$$

$$= -\eta^{ii}({}^3h_{i0,i\,,j} - {}^2h_{ii,0\,,j} + {}^2h_{ij,i\,,0}), \text{ use } i \leftrightarrow j,$$

$$= -\eta^{jj}({}^3h_{j0,j\,,i} - {}^2h_{jj,0\,,i} + {}^2h_{ji,j\,,0}). \tag{8.38}$$

Differentiate Eq. (8.36) with respect to x^k and obtain

$$0 = {}^2h_{00,i\,,k}/2 + \eta^{jj}([\,{}^2h_{ji,j\,,k}\,] - {}^2h_{jj,i\,,k}/2), \text{ use } i \leftrightarrow k,$$

$$= ({}^2h_{00,i\,,k} + \eta^{jj}([\,{}^2h_{ji,j\,,k} + {}^2h_{jk,j\,,i}\,] - {}^2h_{jj,i\,,k})/2. \tag{8.39}$$

Equations (8.38) and (8.39) allow simplified expressions for the Ricci tensor elements,

$$^3R_{0i} = -\nabla^2\,{}^3h_{i0}/2, \quad {}^2R_{ij} = -\nabla^2\,{}^2h_{ij}/2. \tag{8.40}$$

In order to solve Eq. (8.31), the expansion properties of the energy–momentum tensor, studied in Section 7.6, must be examined. Going to an SR frame is the most transparent procedure. $T^{\bar0\bar0}$, $T^{\bar i\bar0}$, and $T^{\bar i\bar j}$ are the energy, momentum, and momentum flux densities, respectively. Thus,

$$T^{\bar0\bar0} = \sum_n P_n{}^{\bar0}\delta(\vec r - \vec r_n(t)) \approx \rho_M(1 + \bar v^2/2 + \cdots)$$

$$= {}^2T^{\bar0\bar0} + {}^4T^{\bar0\bar0} + \cdots, \tag{8.41}$$

$$T^{\bar i\bar0} = \sum_n P_n^{\bar i}\delta(\vec r - \vec r_n(t)) \approx \rho_M\bar v(1 + \bar v^2/2 + \cdots)$$

$$= {}^3T^{\bar i\bar0} + {}^5T^{\bar i\bar0} + \cdots, \tag{8.42}$$

$$T^{\bar i\bar j} = \sum_n P_n^{\bar i}v_n^{\bar j}\delta(\vec r - \vec r_n(t)) \approx \rho_M\bar v^2(1 + \bar v^2/2 + \cdots)$$

$$= {}^4T^{\bar i\bar j} + {}^6T^{\bar i\bar j} + \cdots. \tag{8.43}$$

Even if not in an SR frame, the above are the general expansions for the energy–momentum tensor. Just remove the bars identifying tensor elements. It might seem surprising that the lowest order term in T^{00} is labeled with a "2" on the upper left. It has no explicit factors of $\bar v$, however, mass/volume has units of $M'/V' \propto \bar v^2 r/V'$. This is just what is needed for Eqs. (8.31) and (8.32). In the case of $^2R_{ij}$, the term $^2T^{00}$ comes from the rightmost term in Eq. (8.31). A similar explanation holds for the term $^3T^{i0}$.

So keeping the lowest order term in $T^{\mu\nu}$, and using Eqs. (8.18)–(8.20), (8.31), (8.32), and (8.40) yields

$$-\nabla^2\,{}^2h_{00}/2 = {}^2R_{00} = g_{0\mu}g_{0\nu}\,{}^2R^{\mu\nu} = {}^2R^{00}$$

$$= 8\pi({}^2T^{00} - \eta_{00}g_{\mu\nu}T^{\mu\nu}/2) = 8\pi({}^2T^{00} - (\eta_{00})^2\,{}^2T^{00}/2)$$

$$= 4\pi\,{}^2T^{00} = 4\pi\rho_M, \tag{8.44}$$

$$-\nabla^2\,{}^3h_{i0}/2 = {}^3R_{0i} = g_{0\mu}g_{i\nu}\,{}^3R^{\mu\nu} = \eta_{00}\eta_{ij}\,{}^3R^{0j} = -\,{}^3R^{0i}$$

$$= -8\pi({}^3T^{0i} - g_{0i}g_{\mu\nu}T^{\mu\nu}/2)$$

$$= -8\pi\,{}^3T^{0i}, \tag{8.45}$$

$$-\nabla^2 \, {}^2h_{ij}/2 = {}^2R_{ij} = g_{i\mu}g_{j\nu} \, {}^2R^{\mu\nu} = \eta_{ik}\eta_{jm} \, {}^2R^{km} = {}^2R^{ij}$$

$$= 8\pi(T^{ij} - g_{ij}g_{\mu\nu}T^{\mu\nu}/2) = -8\pi\eta_{ij}\eta_{00} \, {}^2T^{00}/2$$

$$= 4\pi\eta_{ij} \, {}^2T^{00} = 4\pi\eta_{ij}\rho_M. \tag{8.46}$$

The result is that the solutions to Eqs. (8.44) and (8.46) are the same. This is what was found in Chapter 5. The solution that vanishes at $|\vec{r}| = \infty$ is

$$-\frac{{}^2h_{ii}}{2} = -\frac{{}^2h_{00}}{2} = \Psi_G = -\int dV' \frac{{}^2T^{00}(\vec{r}',t)}{|\vec{r} - \vec{r}'|} \approx -\frac{M'}{|\vec{r}|}, \tag{8.47}$$

where the volume integral is over all space. The approximation is for a point external to a spherically symmetric source mass M'. Similarly, the solution for ${}^3h_{0i}$, Eq. (8.45), yields a 3-vector potential $\vec{\Phi}$. Unlike Ψ_G, this potential was unknown before GR, however, like Ψ_G its origin is the GR field equations. Thus,

$$ {}^3h_{0i} = {}^3h^{0i} = \Phi_i = \Phi^i = -4\int dV' \frac{{}^3T^{i0}(\vec{r}',t)}{|\vec{r} - \vec{r}'|}. \tag{8.48}$$

The integral in Eq. 8.48 will now be expanded, and because gravity is so weak, only the lowest order terms are kept,

$$\Phi_i = -4\int dV' \, {}^3T^{i0}(\vec{r}',t)|\vec{r} - \vec{r}'|^{-1}$$

$$= -4\int dV' \, {}^3T^{i0}(\vec{r}',t)(r^2 + r'^2 - 2\vec{r}\cdot\vec{r}')^{-1/2}$$

$$\approx -\frac{4}{r}\left(\int dV' \, {}^3T^{i0}(\vec{r}',t) + \frac{1}{r^2}\eta_{jj}x^j\int dV'x'^j \, {}^3T^{i0}(\vec{r}',t)\right). \tag{8.49}$$

This can be further simplified using energy and momentum conservation,

$$T^{\mu\nu}{}_{;\mu} = 0 \quad \text{so} \quad T^{\mu\nu}{}_{,\mu} = -\Gamma^\nu_{\mu\xi}T^{\mu\xi} - \Gamma^\mu_{\mu\xi}T^{\xi\nu},$$

$$T^{\mu 0}{}_{,\mu} = -\Gamma^0_{\mu\xi}T^{\mu\xi} - \Gamma^\mu_{\mu\xi}T^{\xi 0}.$$

Terms like ΓT are, at minimum, directly proportional to $\bar{v}^4$. They may be neglected. Also $T^{\mu\nu}$ is due to earth and is static. Thus, to the order needed, the above condition yields

$$0 = {}^{2}T^{00}{}_{,0} + {}^{3}T^{j0}{}_{,j}$$

$$= {}^{3}T^{j0}{}_{,j} .$$

This means that the following integrals vanish:

$$0 = \int dV' x'^{i} \; {}^{3}T^{j0}(\vec{r}')_{,j} = \int dV' x'^{i} x'^{j} \; {}^{3}T^{k0}(\vec{r}')_{,k} .$$

Integrate each integral by parts,

$$0 = -\int dV' \; {}^{3}T^{j0}(\vec{r}')x'^{i}{}_{,j} = -\int dV' \; {}^{3}T^{j0}(\vec{r}')\delta^{i}{}_{j}$$

$$= -\int dV' \; {}^{3}T^{i0}(\vec{r}'), \tag{8.50}$$

$$0 = \int dV' x'^{i} x'^{j} \; {}^{3}T^{k0}(\vec{r}')_{,k}$$

$$= -\int dV' (x'^{i} x'^{j}{}_{,k} + x'^{i}{}_{,k} x'^{j}) \; {}^{3}T^{k0}(\vec{r}')$$

$$= -\int dV' [x'^{i} \; {}^{3}T^{j0}(\vec{r}') + x'^{j} \; {}^{3}T^{i0}(\vec{r}')],$$

$$\int dV' x'^{j} \; {}^{3}T^{i0}(\vec{r}') = -\int dV' x'^{i} \; {}^{3}T^{j0}(\vec{r}'), \quad \text{thus,}$$

$$2\int dV' x'^{i} \; {}^{3}T^{j0}(\vec{r}') = \int dV' x'^{i} \; {}^{3}T^{j0}(\vec{r}') - \int dV' x'^{j} \; {}^{3}T^{i0}(\vec{r}'). \tag{8.51}$$

Since ${}^{3}T^{0i}$ is the source momentum density, Eqs. (8.49), (8.50), and (8.51) allow $\vec{\Phi}$ to be expressed in terms of the source's angular momentum $\vec{J} = \vec{r}' \times \vec{P}[\vec{r}']$,

$$\Phi^{1}(\vec{r}) = -\frac{2}{r^{3}} x^{2} \int dV' [x'^{2} \; {}^{3}T^{10}(\vec{r}') - x'^{1} \; {}^{3}T^{20}(\vec{r}')]$$

$$-\frac{2}{r^{3}} x^{3} \int dV' [x'^{3} \; {}^{3}T^{10}(\vec{r}') - x'^{1} \; {}^{3}T^{30}(\vec{r}')]$$

$$= -(2/r^{3})[-x^{2} J^{3} + x^{3} J^{2}] = (2/r^{3})[\vec{r} \times \vec{J}]^{1},$$

$$\vec{\Phi}(\vec{r}) = (2/r^{3})[\vec{r} \times \vec{J}]. \tag{8.52}$$

This is the way earth's angular momentum finds its way into the GP-B predictions.

8.4.4 *Spin*

In SR, a particle has constant velocity and spin,

$$\frac{dU^{\bar{\mu}}}{d\tau} = 0, \quad U^{\bar{\mu}} = \frac{dx^{\bar{\mu}}}{d\tau},$$
$$\frac{dS_{\bar{\mu}}}{d\tau} = 0. \tag{8.53}$$

In the particle's rest frame,

$$S_{\bar{\mu}} \equiv (0, \vec{S}), \quad \text{thus,} \tag{8.54}$$

$$S_{\bar{\mu}} U^{\bar{\mu}} = 0. \tag{8.55}$$

This is an invariant and holds in any frame.

For spin, the equation of motion is obtained from parallel transport Eq. (4.2), with $dq = d\tau$ for a free falling massive object,

$$0 = W^{\nu} S_{\mu;\nu} = \frac{dx^{\nu}}{d\tau}(S_{\mu,\nu} - \Gamma^{\xi}_{\mu\nu} S_{\xi}),$$

$$\frac{dS_{\mu}}{d\tau} = \Gamma^{\xi}_{\mu\nu} S_{\xi} U^{\nu} = \Gamma^{\xi}_{\mu\nu} S_{\xi} \frac{dx^{\nu}}{d\tau},$$

$$\frac{dS_i}{d\tau} = \Gamma^{0}_{i0} S_0 \frac{dx^0}{d\tau} + \Gamma^{k}_{i0} S_k \frac{dx^0}{d\tau} + \Gamma^{0}_{ij} S_0 \frac{dx^j}{d\tau} + \Gamma^{k}_{ij} S_k \frac{dx^j}{d\tau},$$

$$\frac{dS_i}{dt} = \Gamma^{0}_{i0} S_0 + \Gamma^{k}_{i0} S_k + \Gamma^{0}_{ij} S_0 \frac{dx^j}{dt} + \Gamma^{k}_{ij} S_k \frac{dx^j}{dt}. \tag{8.56}$$

The last equation results from multiplying the previous equation by $\frac{d\tau}{dt}$. From Eq. (8.55), with $U^0 \approx 1$ and $v^i \ll 1$,

$$S_0 = -v^i S_i. \tag{8.57}$$

Then to order $\bar{v}^3 \vec{S}$

$$\frac{dS_i}{dt} = -\Gamma^{0}_{i0} v^m S_m + \Gamma^{k}_{i0} S_k - \Gamma^{0}_{ij} S_m v^m v^j + \Gamma^{k}_{ij} S_k v^j$$

$$= [-\,^2\Gamma^{0}_{i0} v^m S_m] + (^3\Gamma^{k}_{i0} S_k) + \,^2\Gamma^{k}_{ij} S_k v^j$$

$$= [\,^2 h_{00,i} v^m S_m/2] + \eta^{kk}((^3 h_{k0,i} - \,^3 h_{i0,k} + \,^2 h_{ki,0}) S_k)/2$$

$$+ \eta^{kk}(^2 h_{ki,j} - \,^2 h_{ij,k} + \,^2 h_{kj,i}) S_k v^j/2)$$

$$= -\Psi_{G,i}\,\vec{v}\cdot\vec{S} + \eta^{kk}(\Phi_{k,i} - \Phi_{i,k} - 2\eta_{ki}\Psi_{G,0})S_k/2$$
$$+ \eta^{kk}(-\eta_{ki}\Psi_{G,j} + \eta_{ij}\Psi_{G,k} - \eta_{kj}\Psi_{G,i})S_k v^j.$$

For the component S_1,

$$\frac{dS_1}{dt} = -\Psi_{G,1}\,\vec{v}\cdot\vec{S} + (\Phi_{2,1} - \Phi_{1,2})S_2/2 + (\Phi_{3,1} - \Phi_{1,3})S_3/2$$

$$- \Psi_{G,0}\,S_1 - S_1\vec{v}\cdot\vec{\nabla}\Psi_G + v^1\vec{S}\cdot\vec{\nabla}\Psi_G - \Psi_{G,1}\,\vec{v}\cdot\vec{S}$$

$$= -\Psi_{G,1}\,\vec{v}\cdot\vec{S} + [(\vec{\nabla}\times\Phi)_3 S_2 - (\vec{\nabla}\times\Phi)_2 S_3]/2$$

$$- \Psi_{G,0}\,S_1 - S_1\vec{v}\cdot\vec{\nabla}\Psi_G + v^1\vec{S}\cdot\vec{\nabla}\Psi_G - \Psi_{G,1}\,\vec{v}\cdot\vec{S}$$

$$= -\Psi_{G,1}\,\vec{v}\cdot\vec{S} + [\vec{S}\times(\vec{\nabla}\times\Phi)/2]_1$$

$$- \Psi_{G,0}\,S_1 - S_1\vec{v}\cdot\vec{\nabla}\Psi_G + v^1\vec{S}\cdot\vec{\nabla}\Psi_G - \Psi_{G,1}\,\vec{v}\cdot\vec{S}.$$

For the full spin vector, it is obvious,

$$\frac{d\vec{S}}{dt} = \vec{S}\times(\vec{\nabla}\times\vec{\Phi})/2 - \vec{S}\Psi_{G,0} - 2\vec{\nabla}\Psi_G(\vec{v}\cdot\vec{S})$$

$$- \vec{S}(\vec{v}\cdot\vec{\nabla}\Psi_G) + \vec{v}(\vec{S}\cdot\vec{\nabla}\Psi_G). \tag{8.58}$$

The invariant $S_\mu S^\mu$ doesn't change,

$$0 = \frac{d(S_\mu S^\mu)}{dt}$$

$$= \frac{d}{dt}(g^{00}S_0 S_0 + g^{i0}S_i S_0) + g^{0j}S_0 S_j + g^{ij}S_i S_j)$$

$$= \frac{d}{dt}(g^{00}v^i S_i v^j S_j - g^{i0}S_i v^j S_j) - g^{0j}v^i S_i S_j + g^{ij}S_i S_j)$$

$$= \frac{d}{dt}(\eta^{00}v^i S_i v^j S_j + [\eta^{ij} + {}^2 h^{ij}]S_i S_j), \text{ use Eq. (8.24)},$$

$$= \frac{d}{dt}(-(\vec{v}\cdot\vec{S})^2 + |\vec{S}|^2 - {}^2 h_{ij}S_i S_j)$$

$$= \frac{d}{dt}(-(\vec{v}\cdot\vec{S})^2 + |\vec{S}|^2 + 2\eta_{ij}\Psi_G S_i S_j)$$

$$= \frac{d}{dt}(-(\vec{v}\cdot\vec{S})^2 + |\vec{S}|^2 + 2\Psi_G|\vec{S}|^2). \tag{8.59}$$

Thus, what is being differentiated is a constant.

Since Ψ_G is of order $\bar{v}^2$, an alternative spin vector correct to order $\bar{v}^2 S$, can be defined,

$$\vec{s} \equiv \vec{S} + \Psi_G \vec{S} - \vec{v}(\vec{v} \cdot \vec{S})/2),$$

$$|\vec{s}|^2 = (1 + \Psi_G)^2 |\vec{S}|^2 + |\vec{v}|^2 (\vec{v} \cdot \vec{S})^2/4 - (1 + \Psi_G)(\vec{v} \cdot \vec{S})^2 \qquad (8.60)$$

$$\approx (1 + 2\Psi_G)|\vec{S}|^2 - (\vec{v} \cdot \vec{S})^2.$$

This result is the same as Eq. (8.59). The solution for $\vec{S}$ in terms of $\vec{s}$, to the same order, is easy to obtain

$$(1 - \Psi_G)\vec{s} = (1 - \Psi_G^2)\vec{S} - (1 - \Psi_G)\vec{v}(\vec{v} \cdot \vec{S})/2)$$

$$\approx \vec{S} - \vec{v}(\vec{v} \cdot \vec{S})/2,$$

$$\vec{S} = (1 - \Psi_G)\vec{s} + \vec{v}(\vec{v} \cdot \vec{S})/2$$

$$\approx (1 - \Psi_G)\vec{s} + \vec{v}(\vec{v} \cdot \vec{s})/2. \qquad (8.61)$$

The vector $\vec{s}$ proves useful if one calculates its total time derivative to order $\bar{v}^3 \vec{S}$. Using $\frac{d\vec{v}}{dt} = -\vec{\nabla}\Psi_G$ and neglecting terms like $\bar{v}\frac{d\vec{S}}{dt}$,

$$\frac{d\vec{s}}{dt} = (1 + \Psi_G)\frac{d\vec{S}}{dt} + \vec{S}\frac{d\Psi_G}{dt} - \left(\frac{d\vec{v}}{dt}\vec{v} \cdot \vec{S} + \vec{v}\left[\vec{S} \cdot \frac{d\vec{v}}{dt} + \vec{v} \cdot \frac{d\vec{S}}{dt}\right]\right) \Big/ 2$$

$$\approx \frac{d\vec{S}}{dt} + \vec{S}\frac{d\Psi_G}{dt} + (\vec{\nabla}\Psi_G[\vec{v} \cdot \vec{S}] + \vec{v}[\vec{\nabla}\Psi_G \cdot \vec{S}])/2$$

$$= \frac{d\vec{S}}{dt} + \vec{S}(\Psi_{G,0} + \vec{\nabla}\Psi_G \cdot \vec{v}) + (\vec{\nabla}\Psi_G[\vec{v} \cdot \vec{S}] + \vec{v}[\vec{\nabla}\Psi_G \cdot \vec{S}])/2.$$

Now substitute Eq. (8.58) for $\frac{d\vec{S}}{dt}$ and Eq. (8.52) for Φ,

$$\frac{d\vec{s}}{dt} = \vec{S} \times (\vec{\nabla} \times \vec{\Phi})/2 - \vec{S}\Psi_{G,0} - 2\vec{\nabla}\Psi_G(\vec{v} \cdot \vec{S}) - \vec{S}(\vec{v} \cdot \vec{\nabla}\Psi_G) + \vec{v}(\vec{S} \cdot \vec{\nabla}\Psi_G)$$

$$+ \vec{S}(\Psi_{G,0} + \vec{\nabla}\Psi_G \cdot \vec{v}) + (\vec{\nabla}\Psi_G[\vec{v} \cdot \vec{S}] + \vec{v}[\vec{\nabla}\Psi_G \cdot \vec{S}])/2$$

$$= -\frac{1}{2}[\vec{\nabla} \times \vec{\Phi} \times \vec{S} + 3(\vec{\nabla}\Psi_G(\vec{v} \cdot \vec{S}) - \vec{v}(\vec{S} \cdot \vec{\nabla}\Psi_G))]$$

$$= -\frac{1}{2}[\vec{\nabla} \times \vec{\Phi} + 3\vec{v} \times \vec{\nabla}\Psi_G] \times \vec{S}, \quad \text{use Eq. (8.52)},$$

$$= -[\vec{\nabla} \times \frac{(\vec{r} \times \vec{J})}{r^3} + \frac{3}{2}\vec{v} \times \vec{\nabla}\Psi_G] \times \vec{S}.$$

Equation (8.58) showed that $\frac{d\vec{S}}{dt}$ was calculated to order $\bar{v}^3 \vec{S}$. The aim is to calculate $\frac{d\vec{s}}{dt}$ to order $\bar{v}^3 \vec{s}$. The terms on the right-hand side of the above equation are to order $\bar{v}^3 \vec{S}$. To obtain the desired order for $\frac{d\vec{s}}{dt}$, use Eq. (8.61) with $\vec{S} \approx \vec{s}$. The above becomes the classical equation for a precessing spin,

$$\frac{d\vec{s}}{dt} = \vec{\Omega} \times \vec{s},$$

$$\vec{\Omega} = -\left[\vec{\nabla} \times \frac{(\vec{r} \times \vec{J})}{r^3} + \frac{3}{2}\vec{v} \times \vec{\nabla}\Psi_G \right]. \qquad (8.62)$$

The spin precesses at a rate $|\vec{\Omega}|$ around the direction of $\vec{\Omega}$ without change in magnitude.

8.4.5 *Numerical Predictions for the GP-B Experiment*

One can show how Eq. (8.62) is used to predict what GP-B expected. The gyroscopes are freely falling in orbit about earth. The origin of coordinates is earth's center and the approximation used for calculating Ψ_G is that the earth is a spherically symmetric mass,

$$\Psi_G = -\frac{M_e}{r}, \quad \vec{\nabla}\Psi_G = M_e\frac{\vec{r}}{r^3}.$$

Then, noting J_e is constant,

$$\vec{\nabla} \times \left(\frac{\vec{r}}{r^3} \times \vec{J}_e \right) = \frac{\vec{r}}{r^3}(\vec{\nabla} \cdot \vec{J}_e) - \vec{J}_e\left(\vec{\nabla} \cdot \frac{\vec{r}}{r^3} \right)$$

$$+ (\vec{J}_e \cdot \vec{\nabla})\frac{\vec{r}}{r^3} - \left(\frac{\vec{r}}{r^3} \cdot \vec{\nabla} \right)\vec{J}_e$$

$$= -\vec{J}_e\left(\vec{\nabla} \cdot \frac{\vec{r}}{r^3} \right) + (\vec{J}_e \cdot \vec{\nabla})\frac{\vec{r}}{r^3}$$

$$= (\vec{J}_e \cdot \vec{\nabla})\frac{\vec{r}}{r^3}, \text{ since } r \neq 0,$$

$$= J_e^x\left(\frac{\vec{r}}{r^3} \right)_{,x} + J_e^y\left(\frac{\vec{r}}{r^3} \right)_{,y} + J_e^z\left(\frac{\vec{r}}{r^3} \right)_{,z}$$

$$= \left[\frac{J_e^x}{r^3} \left(1 - 3\frac{x^2}{r^2}\right) \hat{e}_x - 3\frac{xy}{r^2}\hat{e}_y - 3\frac{xz}{r^2}\hat{e}_z \right]$$

$$+ \left[\frac{J_e^y}{r^3} \left(1 - 3\frac{y^2}{r^2}\right) \hat{e}_y - 3\frac{yx}{r^2}\hat{e}_x - 3\frac{yz}{r^2}\hat{e}_z \right]$$

$$+ \left[\frac{J_e^z}{r^3} \left(1 - 3\frac{z^2}{r^2}\right) \hat{e}_z - 3\frac{zx}{r^2}\hat{e}_x - 3\frac{zy}{r^2}\hat{e}_y \right]$$

$$= \frac{\vec{J}_e}{r^3} - \frac{3\vec{J}_e \cdot \vec{r}}{r^3} \frac{\vec{r}}{r^2}. \tag{8.63}$$

The above form is due to treating $\vec{r}$ as a general 3-vector. You know the GP-B orbit is planar. If this condition was imposed above, by taking $\vec{r}$ as a vector with two components, then $\vec{\nabla} \cdot \vec{r}/r^3 \neq 0$. An incorrect result would have been obtained.

The precession rate is, from Eqs. (8.62) and (8.63),

$$\vec{\Omega} = \vec{\Omega}_{\text{geo}} + \vec{\Omega}_{fd}$$

$$= -\left(\frac{3}{2}M_e \frac{\vec{v} \times \vec{r}}{r^3}\right) + \left(\frac{3\vec{J}_e \cdot \vec{r}}{r^3} \frac{\vec{r}}{r^2} - \frac{\vec{J}_e}{r^3}\right).$$

This formula agrees with Eq. (3) in Schiff's paper.

Figure 1 of the GP-B paper defines the experimental geometry. The gyroscope and its housing are launched in a close in, circular, polar orbit. The orbit rotation is counter clockwise, leading to an angular momentum vector that points out of the orbit plane and toward the reader. This is taken as the $\hat{e}_z$ direction. Earth's angular momentum is directed from south to north pole, and is in the orbit plane pointing up. This is taken as the $\hat{e}_x$-direction. Then, $\hat{e}_y = \hat{e}_z \times \hat{e}_x$ lies in the orbit plane, and is in the direction indicated for the initial spin.

The position of the gyroscope is

$$\vec{r} = \rho\hat{e}_\rho = \rho(\hat{e}_x \cos\phi + \hat{e}_y \sin\phi), \quad \rho = 1.1R_e,$$

where ϕ is the angle between $\vec{J}_e$ and $\vec{r}$. Then,

$$\vec{v} = (M_e/\rho)^{1/2}\hat{e}_z \times \hat{e}_\rho = (M_e/\rho)^{1/2}\hat{e}_\phi,$$

$$M_e = 4.431 \times 10^{-3} \,\text{m}, \ R_e = 6.378 \times 10^3 \,\text{m}, \ T_e = 2.584 \times 10^{13} \,\text{m},$$

$$J_e = 0.829(2M_e R_e^2/5)\omega_e \,\text{m}^2, \ \omega_e = 2\pi/T_e.$$

First, consider the geodetic precession rate,

$$\vec{\Omega}_{\text{geo}} = \frac{3}{2}\left(\frac{M_e}{\rho}\right)^{1/2}\frac{M_e}{\rho^2}\hat{e}_z = \frac{3}{2}\left(\frac{M_e}{\rho}\right)^{3/2}\frac{1}{\rho}\hat{e}_z,$$

$$|\vec{\Omega}_{\text{geo}}| = \frac{1}{3}\times 10^{-20}\,\text{m}^{-1} = 1.02\times 10^{-12}\,\text{s}^{-1}.$$

In a year, $t = 3.156\times 10^7$ s,

$$|\vec{\Omega}_{\text{geo}}|t = 3.22\times 10^{-5}\,\text{rad} = 6.6'',$$

in agreement with the GP-B expectation. The precession direction,

$$\vec{\Omega}_{\text{geo}}\times \hat{e}_y = \hat{e}_z\times \hat{e}_y = -\hat{e}_x,$$

is in the plane of the orbit and downward, in agreement with the first figure in the GP-B paper. This geodetic precession precesses in the same sense as the orbit rotation. It doesn't appear to depend on GR. However, it actually comes from terms in the equations of motion required from curved space, and so it is a consequence of GR.

Now consider the frame-drag precession $\vec{\Omega}_{fd}$. It depends on $\vec{J}_e$,

$$\vec{\Omega}_{fd} = \frac{1}{\rho^3}\left(\frac{3J_e x}{\rho}\hat{e}_\rho - J_e\hat{e}_x\right)$$

$$= \frac{J_e}{\rho^3}(3\cos\phi[\cos\phi\hat{e}_x + \sin\phi\hat{e}_y] - \hat{e}_x).$$

Averaging over a revolution, the $\cos^2\phi$ term yields $1/2$, while the $\cos\phi\sin\phi$ term yields zero. Thus,

$$\langle\vec{\Omega}_{fd}\rangle = \frac{J_e}{\rho^3}(3/2 - 1)\hat{e}_x = \frac{1}{2}\frac{J_e}{\rho^3}\hat{e}_x,$$

$$\frac{|\langle\vec{\Omega}_{fd}\rangle|}{|\vec{\Omega}_{\text{geo}}|} = \frac{1}{3}\frac{J_e}{M_e^{3/2}}\frac{1}{\rho^{1/2}} = 6\times 10^{-3},$$

in agreement with the GP-B expectation. The precession direction,

$$\langle\vec{\Omega}_{fd}\rangle\times \hat{e}_y = \hat{e}_x\times \hat{e}_y = \hat{e}_z,$$

is out of the orbit plane. The spin precesses in the same sense as the earth rotates, hence frame-drag. This agrees with what is shown in the first figure of the GP-B paper.

8.5 Kerr Space Geodesics

8.5.1 *Equations of Motion*

Motion in the equatorial plane $\theta = \pi/2$ of a black hole is considered. Problem 7 shows this is a possible solution, even in Kerr space. The equations of motion for objects with finite rest mass m are most easily obtained by noting $0 = g_{\mu\nu,0} = g_{\mu\nu,2}$. So $U_0 = P_0/m$ and $U_2 = P_2/m$ are constant. The constants will be chosen so that the equations of motion match Eqs. (6.2)–(6.5), with Eq. (6.6), when $a = 0$. Thus,

$$\frac{d\phi}{d\tau} = U^2 = g^{2\nu}U_\nu$$

$$= g^{20}U_0 + g^{22}U_2 = \frac{g_{20}U_0 - g_{00}U_2}{\Lambda}. \tag{8.64}$$

When $a = 0$,

$$\frac{d\phi}{d\tau} = U_2\frac{-g_{00}}{\Lambda} = U_2\frac{(1 - R/r)}{r^2(1 - R/r)} = \frac{U_2}{r^2}.$$

Equations (6.2) and (6.6) yield

$$\frac{d\phi}{d\tau} = \frac{J}{r^2 E'^{1/2}}, \quad \text{so choose,}$$

$$U_2 = \frac{J}{E'^{1/2}}. \tag{8.65}$$

Carrying on,

$$\frac{dt}{d\tau} = U^0 = g^{0\nu}U_\nu$$

$$= g^{00}U_0 + g^{02}U_2 = \frac{-g_{22}U_0 + g_{02}U_2}{\Lambda}. \tag{8.66}$$

When $a = 0$,

$$\frac{dt}{d\tau} = \frac{-g_{22}U_0}{\Lambda} = -\frac{U_0}{1 - R/r}.$$

Equations (6.3) and (6.6) with $2M' = R$ yield

$$\frac{dt}{d\tau} = \frac{1}{E'^{1/2}(1 - R/r)}, \quad \text{so choose,}$$

$$U_0 = -\frac{1}{E'^{1/2}}. \tag{8.67}$$

Using the above constants and the invariant $U_\nu U^\nu$ yields

$$-1 = U_\nu U^\nu = U_0 U^0 + U_2 U^2 + g_{33}(U^3)^2$$

$$= -\frac{g_{22} + 2g_{02}J + g_{00}J^2}{E'\Lambda} + \frac{r^2}{\Lambda}\left(\frac{dr}{d\tau}\right)^2,$$

$$\left(\frac{dr}{d\tau}\right)^2 = -\frac{\Lambda}{r^2} + \frac{g_{22} + 2g_{02}J + g_{00}J^2}{E'r^2}$$

$$= -\frac{r^2 + a^2 - rR}{r^2} + \frac{r^2 + (a^2 - J^2) + (J - a)^2 R/r}{E'r^2}.$$

(8.68)

When $a = 0$,

$$\left(\frac{dr}{d\tau}\right)^2 = -(1 - R/r) + \frac{r^2 - J^2(1 - R/r)}{E'r^2}$$

$$= (1 - R/r)(-1 + [(1 - R/r)^{-1} - (J/r)^2](1/E')).$$

Equations (6.5) and (6.6) with $2M' = R$ yield

$$E'\left(\frac{dr}{d\tau}\right)^2 = (1 - R/r)(-E' - (J/r)^2 + (1 - R/r)^{-1}).$$

So the required equations in Chapter 6 are reproduced with the choices in Eqs. (8.65) and (8.67).

The orbit equation is obtained with Eqs. (8.64) and (8.68). Using $\Lambda = (r - R_+)(r - R_-)$,

$$\frac{d\phi}{dr} = \frac{d\phi}{d\tau}\bigg/\frac{dr}{d\tau}$$

$$= \pm\left[-\frac{(r - R_+)(r - R_-)}{r^2} + \frac{r^2 + (a^2 - J^2) + (J - a)^2 R/r}{E'r^2}\right]^{-1/2}$$

$$\times \frac{aR/r + (1 - R/r)J}{E'^{1/2}(r - R_+)(r - R_-)}$$

$$= \pm\left[-\frac{E'(r - R_+)(r - R_-)}{r^2} + 1 - \frac{J^2 - a^2}{r^2} + \frac{(J - a)^2 R}{r^3}\right]^{-1/2}$$

$$\times \frac{J - (J - a)R/r}{(r - R_+)(r - R_-)}.$$

(8.69)

For photons, $E' = 0$,

$$\frac{d\phi}{dr} = \pm \frac{J - (J-a)R/r}{(r-R_+)(r-R_-)} \left[1 - \frac{J^2 - a^2}{r^2} + \frac{(J-a)^2 R}{r^3} \right]^{-1/2}. \quad (8.70)$$

In some of the calculations below, I have been guided by or have used the parameters of S. Chandrasekhar (1983).

8.5.2   *Circular Motion*

First the circular orbits of photons are considered. It is convenient to calculate $\left(\frac{dr}{dq}\right)^2$ and then take $E' = 0$:

$$
\begin{aligned}
\left(\frac{dr}{dq}\right)^2 &= -E' \frac{\Lambda}{r^2} + 1 - \frac{J^2 - a^2}{r^2} + \frac{(J-a)^2 R}{r^3} \\
&= 1 - \frac{J^2 - a^2}{r^2} + \frac{(J-a)^2 R}{r^3},
\end{aligned}
\quad (8.71)
$$

$$1 = \left(\frac{dr}{dq}\right)^2 + \frac{J^2 - a^2}{r^2} - \frac{(J-a)^2 R}{r^3}. \quad (8.72)$$

Per unit mass, the left-hand side of the above equation is a constant energy and on the right-hand side is a kinetic energy term plus an equivalent gravitational potential energy term $\equiv V_p[r]$. For circular motion $r = R_c$, $J = J_c$, $\frac{dr}{dq} = 0$, and $\frac{dV_p}{dr} = 0$. For stable orbits, $\frac{d^2 V_p}{dr^2} > 0$, while for unstable orbits, $\frac{d^2 V_p}{dr^2} < 0$. In this case,

$$0 = 1 + (J_c - a)^2 R/R_c^3 - (J_c^2 - a^2)/R_c^2, \quad (8.73)$$

$$\left. \frac{dV_p}{dr} \right|_{R_c} = 0 = 3(J_c - a)^2 R/R_c - 2(J_c^2 - a^2), \quad (8.74)$$

$$\left. \frac{d^2 V_p}{dr^2} \right|_{R_c} = -3(J_c - a)^2 R/R_c^5 < 0. \quad (8.75)$$

From Eq. (8.75), it is seen that photon circular orbits are unstable. From Eq. (8.74),

$$R_c = \frac{3R}{2} \frac{(J_c - a)^2}{J_c^2 - a^2} = \frac{3R}{2} \frac{J_c - a}{J_c + a}. \quad (8.76)$$

When $a = 0$, there is only one unstable circular orbit for photons $R_c = 3R/2$.

For direct motion, $a > 0$, while for retrograde motion $a < 0$. Inserting Eq. (8.76) into Eq. (8.73) yields

$$0 = 1 + \frac{(J_c - a)^2(J_c + a)^3 2^3 R}{[3R(J_c - a)]^3} - \frac{(J_c^2 - a^2)(J_c + a)^2 2^2}{[3R(J_c - a)]^2},$$

$$(J_c + a)^3 = \frac{27}{4}R^2(J_c - a) = \frac{27}{4}R^2[(J_c + a) - 2a], \qquad (8.77)$$

$$0 = (J_c + a)^3 - \frac{27}{4}R^2(J_c + a) + \frac{27}{2}aR^2.$$

The following identities allow a transparent solution to the above equation:

$$\cos^3 A = (\cos 3A + 3\cos A)/4,$$

$$\cos 3A = \cos 3(A + [1 \pm 1]\pi/3).$$

Take as solution for Eq. (8.77) the following,

$$(J_c + a) \equiv K \cos A,$$

$$(J_c + a)^3 = K^3(\cos 3A + 3\cos A)/4.$$

Solve for the constant K by first eliminating the $\cos A$ terms. Then solve for $\cos 3A$,

$$3K^3/4 = 27KR^2/4, \quad K = \pm 3R,$$

$$\cos 3A = \mp 2a/R, \quad \text{so can take,}$$

$$\cos 3A = 2|a|/R, \qquad (8.78)$$

$$J_c + a = -3R(a/|a|)\cos(A + [1 + a/|a|]\pi/3). \qquad (8.79)$$

When $a = 0$, take $a/|a| = 1$.

For the cases of minimum and maximum $|a|$, the results are as follows:

$$a = 0, \ 3A = \pi/2, \ J_c = 3^{3/2}R/2, \ R_c = 3R/2, \ R_{(+,-)} = (R, 0),$$

$$a = R/2, \ 3A = 0, J_c = R, \ R_c = R/2, \ R_\pm = R/2,$$

$$a = -R/2, \ 3A = 0, \ J_c = 7R/2, \ R_c = 2R, \ R_\pm = R/2.$$

The result for $a = 0$ agrees with the answer to Problem 6.10. The results for J_c/R' and R_c/R', where $R' = R/2$, for all values of a, are shown in Fig. 8.3. The retrograde orbits are farther out, where frame-drag is not so strong.

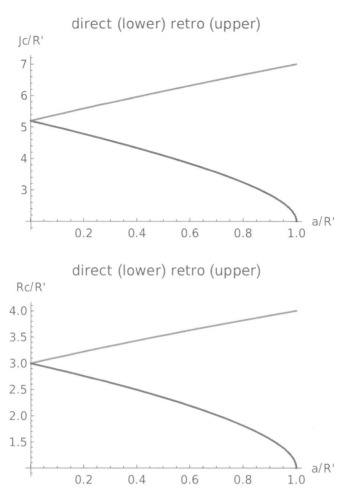

Fig. 8.3 Photon unstable circular orbit parameters for direct and retrograde motion: J_c/R' upper, R_c/R' lower, as functions of a/R' where $R' = R/2$.

Now circular motion of massive objects is studied. Recall such motion was considered in Problems 6.7–10 for the case $a = 0$. Here all values of a are considered. Rewriting Eq. (8.71) yields,

$$\frac{1}{E'} = \left(\frac{dr}{d\tau}\right)^2 + \frac{(r - R_+)(r - R_-)}{r^2}$$
$$+ \frac{1}{E'}\left(\frac{J^2 - a^2}{r^2} - \frac{(J - a)^2 R}{r^3}\right),$$

$$\frac{1}{E'} \equiv \left(\frac{dr}{d\tau}\right)^2 + V[r],$$

$$V = \frac{(r - R_+)(r - R_-)}{r^2} + \frac{1}{E'}\left(\frac{J^2 - a^2}{r^2} - \frac{(J - a)^2 R}{r^3}\right). \qquad (8.80)$$

Once again, per unit mass, there is a constant energy on the left-hand side of Eq. (8.80), and on the right-hand side a kinetic energy term and a gravitational potential energy term $\equiv V$ are present.

The calculations for the circular orbits are given below. At this point, some of those results are used to have a qualitative look at V as shown in Fig. 8.4. When there is a circular orbit, $V = E'^{-1}$. Using $R' = R/2$, plots show V versus r/R' for $a/R' = 0.8$. The top plot uses $E'^{-1} = 0.77$ and $J/R' = 2.71$, which produce a stable circular orbit with minimum radial coordinate $R_c/R' = 2.91$. Note that the minimum just barely exists. In the middle plot $R_c/R' = 4$, and the constants are $E'^{-1} = 0.8$ and $J/R' = 2.8$. Here, the minimum is clear. In the bottom plot, the constant $E'^{-1} = 0.81$ has been changed slightly, while J has not been changed. Due to this small change in E'^{-1}, there is no longer a circular orbit, with minimum at $V = E'^{-1}$. For this case, the orbit is an ellipse with extreme radial coordinates at $r/R' = (3, 5.5)$. Recall, in Chapter 6 the ellipses were shown to be precessing. So qualitatively one sees how the potential determines the bound orbits.

For circular motion with $r = R_c$, $0 = \frac{dr}{d\tau} = \frac{dV}{dr}$,

$$E' = \frac{R_c^2 + (J - a)^2 R/R_c - (J^2 - a^2)}{(R_c - R_+)(R_c - R_-)}, \qquad (8.81)$$

and

$$E' = -\frac{3(J - a)^2 R/2 - (J^2 - a^2)R_c}{R_c(RR_c/2 - a^2)}. \qquad (8.82)$$

In order to solve for $R_c = f[a, R, J, E']$, one must satisfy simultaneously a cubic and a quadratic equation. I have not heard of such a solution.

If $a = 0$, the solution is trivial. Therefore, Eq. (8.82) becomes, with $J^2/E' = (U_2)^2$,

$$0 = R_c^2 - \frac{2(U_2)^2}{R}R_c + 3(U_2)^2,$$

$$R_c = \frac{(U_2)^2}{R}\left(1 \pm \left[1 - \frac{3R^2}{(U_2)^2}\right]^{1/2}\right), \qquad (8.83)$$

$$(U_2)^2 \geq 3R^2. \qquad (8.84)$$

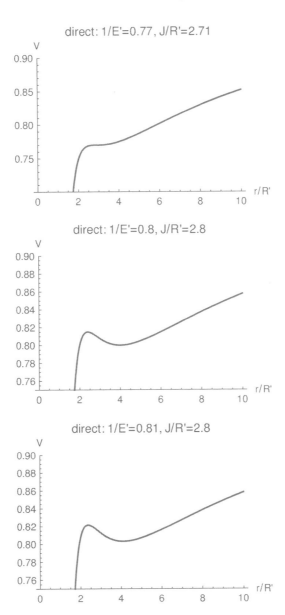

Fig. 8.4   Potential for direct orbits versus r/R' for $a/R' = 0.8$, where $R' = R/2$. Top: E'^{-1}, J that yield the circular orbit with the minimum radial coordinate, Middle: E'^{-1}, J that yield a circular orbit with a larger radial coordinate, Bottom: E'^{-1} changed slightly from the middle plot value, to give a bound, precessing elliptical orbit. Here, the extreme radial coordinates are $(3, \ 5.5)R'$.

This solution is available because two of the four parameters are eliminated. Only a simple quadratic equation needs solving.

The condition for orbit stability is easily found,

$$\frac{d^2V}{dr^2}\bigg|_{R_c} = \frac{2(U_2)^2}{R_c^4}\left(1 - \frac{3R}{R_c}\right) \geq 0. \tag{8.85}$$

So the orbit is stable if $R_c \geq 3R$. This occurs if the positive sign is taken in Eq. (8.83). The minimum stable, circular orbit occurs for,

$$(U_2)^2 = 3R^2, \quad R_c(\text{min}) = 3R. \tag{8.86}$$

This value for $R_c(\text{min})$ is twice the value of the photon circular orbit when $a = 0$. If the negative sign is taken in Eq. (8.83), the minimum unstable, circular orbit occurs for,

$$U_2 \to \infty, \quad R_c(\text{min}) = 3R/2.$$

This is the same as for the photon minimum circular orbit. So here the speed would approach unity. For stable orbits, $R_c \geq 3R$, while for unstable orbits, $3R/2 \leq R_c < 3R$.

However, for any a, if you assume a value for R_c has been found, then solutions for J and E' can be found. These lead to further progress. Equating (8.81) and (8.82) leads to a quadratic equation for J. It is convenient to set $R_c \equiv 1/u$,

$$\frac{1 + (J-a)^2 Ru^3 - (J^2 - a^2)u^2}{1 + a^2u^2 - Ru} = -u\frac{3(J-a)^2 Ru/2 - (J^2 - a^2)}{(R/2 - a^2u)},$$

$$0 = A'J^2 + B'J + C', \text{ where,}$$

$$A' = u[1 - 2uR + u^2R^2 - a^2Ru^3/2]),$$

$$B' = au^2R(3 + a^2u^2 - 2uR),$$

$$C' = -(R/2)(1 + 2u^2a^2 - 2u^3a^2R + u^4a^4),$$

$$J = -\frac{au^2R/2(3 + a^2u^2 - 2uR) \mp (uR/2)^{1/2}(1 + a^2u^2 - uR)}{u[1 - 2uR + u^2R^2 - a^2u^3R/2]}$$

$$= -\frac{auR/2(3 + a^2u^2 - 2uR) \mp (R/[2u])^{1/2}(1 + a^2u^2 - uR)}{(1 - Ru + [a^2u^3R/2]^{1/2})(1 - Ru - [a^2u^3R/2]^{1/2})}. \tag{8.87}$$

Once J is known, Eqs. (8.81) and (8.82) can be treated like a pair of coupled equations. They can be appropriately combined to eliminate the J^2

terms, and allow one to solve for E' in terms of J. Then one uses Eq. (8.67) to get U_0,

$$E' = \frac{(1 - 3uR/2 \mp 2[a^2u^3R/2]^{1/2})(1 - uR \pm [a^2u^3R/2]^{1/2})^2}{(1 - 2uR + u^2R^2 - a^2u^3R/2)^2}$$

$$= \frac{(1 - 3uR/2 \mp 2[a^2u^3R/2]^{1/2})}{(1 - uR \mp [a^2u^3R/2]^{1/2})^2}, \qquad (8.88)$$

$$-U_0 = \frac{1}{E'^{1/2}} = \frac{1 - uR \mp [a^2u^3R/2]^{1/2}}{(1 - 3uR/2 \mp 2[a^2u^3R/2]^{1/2})^{1/2}}.$$

These results check with those of Chandrasekhar. His parameters are $E = E'^{-1/2} = -U_0$ and $L = J/E'^{1/2} = U_2$. The easiest way to verify that the above expressions are the same is start with his expression for L,

$$L = \frac{J}{E'^{1/2}} = \mp \left(\frac{R}{2u}\right)^{1/2} \frac{1 + a^2u^2 \pm (a^2u^3R/2)^{1/2}}{(1 - 3uR/2 \mp 2[a^2u^3R/2]^{1/2})^{1/2}},$$

$$J = LE'^{1/2} = \mp \left(\frac{R}{2u}\right)^{1/2} \frac{1 + a^2u^2 \pm (a^2u^3R/2)^{1/2}}{1 - uR \mp [a^2u^3R/2]^{1/2}} \qquad (8.89)$$

$$= \mp \left(\frac{R}{2u}\right)^{1/2} \frac{1 + a^2u^2 \pm (a^2u^3R/2)^{1/2}}{1 - uR \mp [a^2u^3R/2]^{1/2}} \frac{1 - uR \pm [a^2u^3R/2]^{1/2}}{1 - uR \pm [a^2u^3R/2]^{1/2}}.$$

The reader can confirm the above expression is identical to Eq. (8.87) by multiplying out the numerator, and noting the denominator is the same.

One case of interest is that of marginally bound objects. They have no radial velocity, when very far from the origin. Here $E'^{-1} = V$ and from Eq. (8.80) $V = 1$. The circular orbit radial coordinates $R_{\text{cmb}} \equiv 1/u$ are from Eq. (8.88):

$$[1 - uR \mp (a^2u^3R/2)^{1/2}]^2 = 1 - 3uR/2 \mp 2(a^2u^3R/2)^{1/2}.$$

Thus,

$$0 = (uR/2)[a^2u^2 + 2uR - 1 \pm 4(a^2u^3R/2)^{1/2}],$$

$$1 = a^2u^2 + 2uR \pm 4(a^2u^3R/2)^{1/2} = [au \pm 2(Ru/2)^{1/2}]^2,$$

$$\pm 1 = au \pm 2(Ru/2)^{1/2},$$

$$2Ru = (\pm 1 - au)^2 = 1 \mp 2au + a^2u^2,$$

$$u = [R \pm a \pm (R^2 \pm 2aR)^{1/2}]/a^2,$$

$$R_{\text{cmb}} = \frac{a^2}{R \pm a \pm (R^2 \pm 2aR)^{1/2}} \frac{R \pm a \mp (R^2 \pm 2aR)^{1/2}}{R \pm a \mp (R^2 \pm 2aR)^{1/2}}$$

$$= R \pm a + (R^2 \pm 2aR)^{1/2}. \tag{8.90}$$

The positive sign before the square root must be taken, or the radial coordinate is inside or on the horizon.

It's important to note that, for the remaining $\pm$ sign, the larger R_{cmb} goes with the upper or plus sign, and thus must be for the retrograde orbits. The smaller R_{cmb} goes with the lower or minus sign, and thus must be for the direct orbits. This upper–lower convention must then hold in Eqs. (8.87)–(8.89). Some specific results are as follows:

$$a = 0, \quad R_{\text{cmb}} = 2R,$$

$$a = R/2, \quad R_{\text{cmb}} = R/2 \text{ direct}, \ R(3/2 + 2^{1/2}) \text{ retrograde}.$$

Compare these results, with those obtained for photons, and massive particles with $a = 0$. Photons have the smallest radial coordinate and the massive particles the largest. The direct orbits of marginally bound orbits are in between. At $a = R/2$, all direct orbit radial coordinates are at $R/2$. This will be shown below for the innermost, massive particle, stable, circular orbits. Thus, the graph of R_{cmb} as a function of a will be similar to what is seen in Fig. 8.3, but above those curves. The innermost stable radial coordinates of massive particles would be higher still. As $R_{\text{cmb}} < R_c(\text{min})$, the minimally bound circular orbits are unstable. Forcing $E' = 1$ is the culprit.

The innermost stable radial coordinate of a massive particle occurs when

$$\frac{dE'^{-1}}{du} = -\frac{1}{E'^2} \frac{dE'}{du} = 0.$$

For the case $a = 0$, Eq. (8.82) yields

$$E' \to \frac{2J^2 u}{R} \left(1 - \frac{3Ru}{2}\right),$$

$$0 = \frac{dE'}{du} = 1 - 3Ru,$$

$$R_c(\text{min}) = 1/u = 3R,$$

the expected result. In the general case, Eq. (8.88) yields

$$0 = \frac{dE'^{-1}}{du} = 1 - 3Ru - 3a^2u^2 \mp 8a(R/2)^{1/2}u^{3/2}.$$

This equation can be put in standard form using $u = 1/R_c(\text{min})$, $x = R_c(\text{min})/(R/2)$, and $\sigma = a/(R/2)$. Then,

$$0 = x^2 - 6x \mp 8\sigma x^{1/2} - 3\sigma^2. \tag{8.91}$$

When $\sigma = 1$, the solution for the direct orbit case is $x = 1$, as expected. When $\sigma = 0$, the solution is $x = 6$, as expected. Equation (8.91) has an analytic solution, see https://duetosymmetry.com/tool/kerr-isco-calculator/,

$$f \equiv 1 + (1 - \sigma^2)^{1/3}[(1 + \sigma)^{1/3} + (1 - \sigma)^{1/3}],$$
$$g \equiv (3\sigma^2 + f^2)^{1/2}, \tag{8.92}$$
$$x = 3 + g \pm [(3 - f)(3 + f + 2g)]^{1/2}.$$

Rather than analytically check that Eq. (8.92) is the solution to Eq. (8.91), a numerical evaluation with Mathematica indicates that the value of the latter is everywhere less than 10^{-11}, when the former is used for x.

The result for x yields the innermost radial coordinate for any a. Note that the upper sign yields the larger innermost R_c. It represents retrograde orbits, where the frame-drag is not so potent. The lower sign yields the smaller innermost R_c. It represents direct circular orbits, where frame-drag is strong. Once $R_c(\text{min})$ is determined, E', J, and $\frac{d^2V}{dr^2}$ can be calculated. As for stability, Mathematica indicates,

$$\frac{d^2V}{dr^2}\Big|_{r=R_c(\text{min})} > 0,$$

everywhere, for both direct and retrograde orbits. So V is a minimum and the orbits are stable.

8.5.3 *General Geodesics*

The general geodesics for photons can be viewed schematically using Eq. (8.70). Very interesting orbits occur when $J = J_c$, the value allowing circular orbits of $r = R_c$, and the light starts towards the black hole from $r > R_c$. Equation (8.70) can be rewritten with the aid of Eq. (8.76), to transparently show the importance of R_c,

$$
\begin{aligned}
\pm\frac{d\phi}{dr} &= -\frac{-J_c + (2R'/r)(J_c - a)}{(r - R_+)(r - R_-)}\frac{1}{[1 + (2R'/r^3)(J_c - a)^2 - (J_c^2 - a^2)/r^2]^{1/2}} \\
&= \frac{R_c}{3}\frac{-3J_c/R_c + 2(J_c + a)/r}{(r - R_+)(r - R_-)} \\
&\quad \times \frac{1}{R'^{1/2}(J_c - a)[2/r^3 - 3/R_c/r^2 + ((J_c - a)^2 R')^{-1}]^{1/2}} \\
&= \frac{R_c}{3}\frac{-3J_c/R_c + 2(J_c + a)/r}{(r - R_+)(r - R_-)} \\
&\quad \times \frac{1}{R'^{1/2}(J_c - a)[2/r^3 - 3/R_c/r^2 + 1/R_c^3]^{1/2}} \\
&= \frac{R_c}{3}\frac{-3J_c/R_c + 2(J_c + a)/r}{(r - R_+)(r - R_-)} \\
&\quad \times \frac{1}{R'^{1/2}(J_c - a)(2/r + 1/R_c)^{1/2}(1/r - 1/R_c)} \\
&= \frac{R'^{1/2}}{(J_c + a)}\frac{-3J_c/R_c + 2(J_c + a)/r}{(r - R_+)(r - R_-)}\frac{1}{(2/r + 1/R_c)^{1/2}(1/r - 1/R_c)}.
\end{aligned}
$$
$$(8.93)$$

In Figs. 8.5 and 8.6, $a = \pm 0.4R$ is used, in order to check these results with those of Chandrasekhar. In Fig. 8.6, the orbits are sketched. Note, the circles are not drawn to scale, so the photon path is more clearly shown. The spin of the black hole points out of the page. In order to start the direct, retrograde orbit in the same (counter clockwise), opposite (clockwise) sense as the spin, the derivatives, $\pm\frac{d\phi}{dr}$, shown to scale in Fig. 8.5, require opposite signs. The magnitude of the derivative becomes infinite as $r \to R_c$ from outside. It spends an indeterminate amount of time moving in this circle and the horizon circles, but some instability will force the orbit to smaller r and eventually to the singularity.

In the case of direct motion, the derivative doesn't change sign as R_c is traversed. The light travel continues in the same sense. The derivative decreases in magnitude, and then increases to infinite magnitude, when $r \to R_+$. So the direct photon goes around the horizon, in the same sense as the spin, as you would expect from frame-drag.

Punching through R_+, the derivative discontinuously changes sign, and decreases in magnitude. It passes through zero, where it continuously

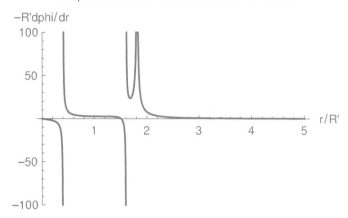

-R'dphi/dr direct versus r/R' for a=0.8R'

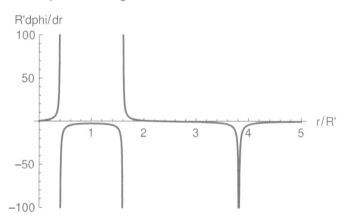

R'dphi/dr retrograde versus r/R' for a=-0.8R'

Fig. 8.5   Photon direct (upper) and retrograde (lower) orbit derivatives $\pm R' \frac{d\phi}{dr}$ for $a =$ $0.8R'$, $R' = R/2$. Other parameters are listed in Fig. 8.6. The opposite signs have been used to start the orbits direct (counter clockwise), retrograde (clockwise), in the same, opposite sense as the black hole spin. The photon approaches from $r > R_c$.

changes sign. It becomes infinitely large as R_- is approached, and so goes around in the same sense as the spin. On punching through R_-, the derivative discontinuously changes sign for the final time. It then decreases in magnitude as the singularity is approached.

For retrograde motion, the derivative magnitude decreases rapidly as R_c is traversed. It crosses zero, continuously changes sign, and increases as

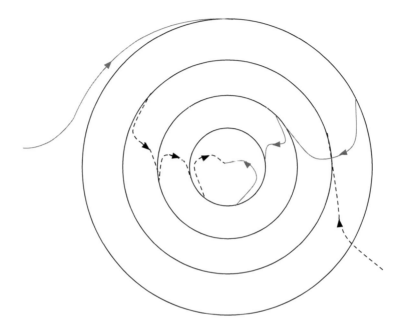

Fig. 8.6   Photon orbit sketch for $a = 0.8R'$, $R' = R/2$. The circles are not drawn to scale to clearly show the photon path. The inner two circles are the horizons $R_{(-,+)} = (0.4, 1.6)R'$. The spin of the black hole points out of the page. The dashed, direct orbit is started in the same sense as the spin, with $J = J_c = 3.237R'$. This yields $R_c = 1.811R'$, the third circle from the center. The solid, retrograde orbit is started in the opposite sense as the spin, with $J = J_c = 6.662R'$. This yields $R_c = 3.819R'$, the biggest circle. The photon approaches from $r > R_c$.

R_+ is approached. It goes around the horizon in the same sense as the spin. This shows the influence of frame-drag, which is extremely strong at the horizon. Once sucked inside R_+, the sign of the derivative discontinuously changes, and the magnitude decreases. It then increases to infinity, as R_- is approached. Traversing R_-, there is a last discontinuous sign change, and the magnitude decreases, as the light proceeds to the singularity.

This very odd behavior for light is predicted by the GR equations. At first, it seems non-scientific, as it cannot be tested. Orbiting photons can't be observed. However, in principle it could be tested. The photon's energy gives rise to a gravitational effect.

The interested reader may enjoy seeing what orbits occur when Eq. (8.70), with different parameters, is input to Mathematica. Once a figure like Fig. 8.5 is obtained, the orbit can be sketched. Of course, the option for numerical integration always exits, but the above procedure is relatively simple. The same games can be played with massive particles using Eq. (8.69).

8.6   An Interstellar Example

The movie "Interstellar" illustrates some wonderful effects concerning GR and black holes. One such effect, showing a distinct difference between a spinning, and a static black hole, is discussed. Astronauts, from a dying earth, punch through a worm hole. They find themselves in a different part of the universe, that otherwise is unreachable, due to the large proper distance from earth. They explore the planets of a solar system, whose star is a huge rotating black hole, aptly named Gargantua. They hope to find a planet that can serve as a new home. One planet is in the minimum, stable, circular orbit, almost as close as possible to the event horizon. This is possible because Gargantua is rotating extremely close to its maximum value $a = (1 - 1.3 \times 10^{-14})R/2$ (Thorne, 2014). An enormous time dilation factor of $\approx 6 \times 10^4$, as required by the movie director, is experienced. The pilot astronaut spends a short time — hour(s) — on the planet, and after the entire trip, has hardly aged. However, when he returns to earth, he finds the young daughter he left behind, is now a very elderly woman.

This example shows how such phenomena can be possible with a spinning, but not a static black hole. The radial coordinate of the planet's stable circular orbit is $r = R_c(\min)$. In Section 8.5, the following was found:

$$R_c(\min) = \begin{cases} 3R, & a = 0, \\ R/2, & a = R/2. \end{cases}$$

Thus, for a static black hole, Eqs. (8.65)–(8.67) and (8.88) yield

$$\frac{1}{E'^{1/2}} = \left(\frac{8}{9}\right)^{1/2},$$

$$\frac{dt}{d\tau} = \frac{3}{2}\left(\frac{8}{9}\right)^{1/2} = 1.4.$$

The time period dt is that read on a faraway at-rest clock, and $d\tau$ is the time period on a clock attached to the planet. So from a static black hole, a time dilation factor at most 1.4 is possible.

In Gargantua's case, you might expect a larger time dilation. The radial coordinate of the event horizon is half, what it would be, in the static case. However, a time dilation factor of 6×10^4 is truly impressive. The innermost, stable circular orbit solutions come from solving Eq. (8.91) and Eq. (8.92) is the exact solution. The exact solution could be evaluated with Mathematica, but it isn't very instructive. Chandrasekhar quotes, unfortunately without proof, the following result:

$$R_c(\text{min}) = (1 + (4\delta)^{1/3})R/2, \quad \text{when } a = \frac{R}{2}(1 - \delta), \ \delta \to 0.$$

Here the proof is indicated. Start with Eq. (8.91), and note $x = R_c(\text{min})/(R/2)$ and $\sigma = a/(R/2)$. If $a = (R/2)(1 - \delta)$, then $x = 1 + \epsilon$, $\epsilon \to 0$. So the differential equation to solve for the direct case is

$$0 = (1 + \epsilon)^2 - 6(1 + \epsilon) + 8(1 - \delta)(1 + \epsilon)^{1/2} - 3(1 - \delta)^2.$$

Now expand $(1 + \epsilon)^{1/2}$ for very small ϵ. Given Chandrasekhar's solution, the expansion should go, at least, to the third order. Thus,

$$0 = 1 + 2\epsilon + \epsilon^2 - 6(1 + \epsilon) + 8(1 - \delta)(1 + \epsilon/2 - \epsilon^2/8 + \epsilon^3/16) - 3(1 - 2\delta)$$

$$= (1 - 6 + 8 - 3) + \epsilon(2 - 6 + 4) + \epsilon^2(1 - 1) + \epsilon^3/2$$

$$+ \delta(-8 + 6 - 4\epsilon + \epsilon^2 - \epsilon^3/2)$$

$$\approx \epsilon^3/2 - 2\delta, \quad \text{thus } \epsilon = (4\delta)^{1/3}.$$

Obviously expansion beyond third order can be neglected. This yields

$$R_c(\text{min}) = (1 + (4\delta)^{1/3})R/2.$$

If one now uses $\delta = 1.3 \times 10^{-14}$, the Thorne value, the following results are obtained:

$$R_c(\text{min}) = 0.500019R, \quad E, \ L = 0.577372 \times (1, \ R), \quad \frac{dt}{d\tau} = 61874.$$

So the movie illustrates a valid scientific effect. The radial coordinate of the planet is extremely close to that of the horizon. However, the calculation of the proper distance from the horizon, left as a problem, indicates it isn't as close to the horizon as you might think. The curvature of space is very large in this region.

Problems

1. A massive, spherically symmetric star goes supernova. The remains of the star are a mass three times that of the sun $M' = 3M_s$ that will collapse radially to a static black hole. If the remains start with a surface

radial coordinate twice that of the sun $r = 2R_s$, what is R/r, where R is the Schwarzschild radius? The surface starts its collapse from rest. As the star collapses, excited ^{4}He atoms at the surface emit radially, photons of characteristic frequency ν', while the atoms continue to freely fall with the rest of the star. What frequency ν will be observed by a stationary observer at $2R_s$ in terms of the photon emission radial coordinate r? When the collapse goes through R show that the photons are red shifted to the extreme. As a hint, recall Eqs. (2.27), (2.28), and (6.2)–(6.6).

2. Other metrics can be used to illustrate that light cannot escape from a singularity. The following metric from the element of proper time will also do the trick,

$$(d\tau)^2 = (1 - M'/r)^2(dt)^2 - (1 - M'/r)^{-2}(dr)^2$$
$$- r^2((d\theta)^2 + \sin^2\theta(d\phi)^2).$$

There are singularities at $r = M'$ and $r = 0$. Change coordinates to (v, r, θ, ϕ) with the transformation, $t = v - f(r)$, so that $g_{rr} = 0$. Then the singularity at $r = M'$ is removed. Find $f_{,r}$, the new metric, and f such that $v = 0$ at $r = t = 0$. Does this metric describe a black hole? That is, can light originating at $r < M'$ get to $r > M'$ while light originating at $r > M'$ can get to both $r < M'$ and $r = \infty$.

3. A particle crosses the event horizon $r = R$ of a Schwarzschild black hole. What is the maximum proper time to get to the singularity at $r = 0$? Suppose the particle starts from rest at $r = 5R$. How much proper time passes before it reaches $r = R$?

4. In Problem 3, the proper time for travel to the horizon of the black hole is finite. Here, the time t on a clock of a faraway at-rest observer is considered. Take a particle at rest far from the black hole so that $E' \approx 1$. Calculate how long it takes to get to the horizon of the black hole?

5. Show Eqs. (8.8) and (8.9) lead to Eqs. (8.10)–(8.12).

6. Compare the ratios of the radial coordinates and areas of the horizon of a static Schwarzschild black hole to a rotating black hole.

7. For the static Schwarzschild black hole, a solution $\theta = \pi/2$ is possible. Write out the equation of motion for the coordinate θ, and show it also is a solution for a rotating black hole?

8. Suppose the orbit of the GP-B gyroscopes was about the equator instead of about the poles, and their orbital angular momentum was

in the same direction as the earth's angular momentum. Calculate the magnitudes and directions of the geodetic and frame-drag precessions. You should find something interesting for the frame-drag precession. Have you any explanation? That is found in Schiff (1960). Explain why a polar orbit is better for the experimenters.

9. Suppose a black hole has maximum spin $a = R/2$. A massive particle with $\theta = \pi/2$ is released from rest at $r = 5R$. What is $\frac{d\phi}{dt}$ and $\frac{dr}{dt}$ at the outer edge of the ergosphere and at the horizon?

10. What is R for the black hole in M87 using its given mass? Do you think a is similar to that of Gargantua or close to the static value? Explain and obtain an estimate for a.

11. Suppose the black hole in M87 was Gargantua and suppose you were on the planet in the closest, stable, circular orbit. The planet is not as close to Gargantua as its radial coordinate implies. What is the proper distance to the horizon? What angle is subtended between Gargantua's center and horizon? Compare this with the angle subtended between the sun's center and its radial edge, as seen from earth. Try to visualize what the view of the sky would be, since you know what the view of the sun is from earth. Gargantua didn't look this big in the movie because the director wanted a different visual effect.

Chapter 9

Cosmology

9.1 Robertson–Walker Metric

There are a large number of particles that group into various structures in the universe. In order to make headway, the cosmological principle is used. This principle states that no position is favored, or every observer's view is the same. At first this seems crazy. Our position is certain to give a view of the solar system that is different from that of the sun or another star. So the proviso, *when averaged over a large enough distance* $\approx 10^9$ pc, is added to the principle. Such a distance includes many galaxy clusters. On this scale, as shown in Fig. 9.1, the universe is spherically symmetric about any point or is isotropic, and looks the same in all directions or is homogeneous. Also when a faraway source is observed, the observer is looking back in time because the radiation travels with finite velocity.

The best model of the universe has it starting in a hot, dense state about 13.8 Bya. It has since been expanding and thus cooling, review Problem 2.15. There is a relic signal from the past, from about 0.37 My after the start, called the cosmic microwave background (CMB). This background, observed in all directions, yields the most ideal Planck black-body spectrum yet found. The CMB originates from the time when the radiation was much hotter, but still cool enough so that atoms could form. At that point, the electrons could no longer scatter the radiation. The universe became transparent to the CMB photons, and they too cooled with the universal expansion. The decreasing temperature is a scalar providing a time

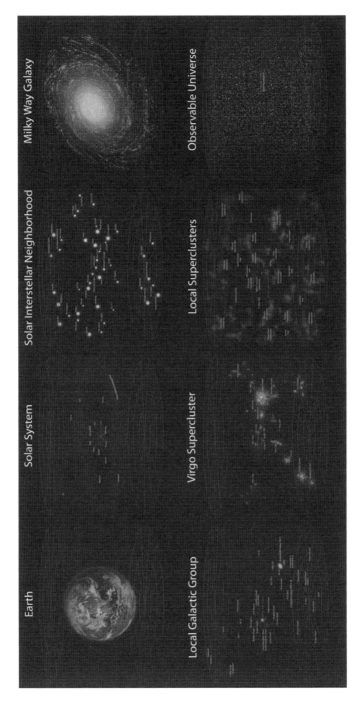

Fig. 9.1 Views of the universe at various distance scales. At the largest scale, the universe appears homogeneous and isotropic. Credit: Andrew Z. Colvin.

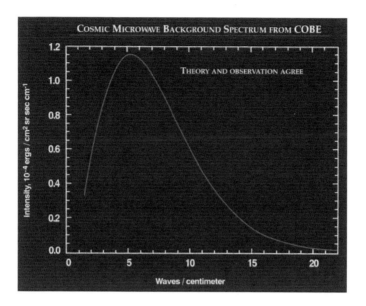

Fig. 9.2 The COBE (2015) Project CMB data compared with a Planck spectrum.

marker for the expansion. In Fig. 9.2, the results of the COBE experiment, available at COBE (2015), is compared with a Planck spectrum. A summary of all the data (Fixsen, 2009) shows that a $T = 2.726\,\mathrm{K}$ spectrum is observed everywhere you look and substantiates a homogeneous, isotropic universe. The theoretical prediction is wider than the error bars on the data points. The COBE, WMAP (2015) and PLANCK (2015) experiments have observed subtle structure in different directions of the sky, on a much finer temperature scale $\approx 10^{-5}\,\mathrm{K}$. This structure is thought to be due to random fluctuations in the extremely young universe, and led to formation of structures observed today.

Experiments are carried out and they yield $g_{\mu\nu}(x^\chi)$, $T_{\mu\nu}(x^\chi)$. A different coordinate system $x^{\chi'}$ is considered equivalent if all of universal history appears the same when expressed in terms of these coordinates. That is, at any place and time, expressed in terms of X^χ, the relations between the two systems are as follows:

$$g_{\mu\nu}(X^\chi) = g_{\mu'\nu'}(X^\chi),$$
$$T_{\mu\nu}(X^\chi) = T_{\mu'\nu'}(X^\chi).$$

Then as cosmic time has been determined from a scalar, that scalar is only a function of time,

$$S(x^0) = S(x^{0'}),$$
$$x^0 = x^{0'}.$$

So all "good" coordinate systems use the same "cosmic" time. These assumptions allow a suitable metric to be obtained.

The expanding universe or cause of the decrease in the universal black-body temperature requires a time-dependent metric of the form,

$$d\tau^2 \equiv dt^2 - Q^2 C_{ij} dx^i dx^j, \quad Q = Q[t]. \tag{9.1}$$

Here, $\vec{r} = x^i \hat{e}_i$, $d\vec{r} = dx^i \hat{e}_i$, and the scale factor Q is only a function of cosmic time. Since there is no preferred direction, $g_{0i} = 0$ and $g_{00} = -1$ because cosmic time is the same for all observers. The quantity C_{ij} is required to make the universe spherically symmetric and seen to be the same if one translates from one origin to any other. The following, with $r^p = (r)^p$, is the simplest form that will do the trick,

$$C_{ij} dx^i dx^j = \eta_{ij} dx^i dx^j + \frac{k(\eta_{mn} x^m dx^n)^2}{1 - k\eta_{st} x^s x^t} = d\vec{r} \cdot d\vec{r} + \frac{k(\vec{r} \cdot d\vec{r})^2}{1 - k\vec{r} \cdot \vec{r}}, \tag{9.2}$$
$$d\vec{r} \cdot d\vec{r} = (dr)^2 + r^2((d\theta)^2 + \sin^2\theta(d\phi)^2), \quad \vec{r} \cdot d\vec{r} = rdr,$$

$$C_{ij} dx^i dx^j = \frac{(dr)^2}{1 - kr^2} + r^2((d\theta)^2 + \sin^2\theta(d\phi)^2). \tag{9.3}$$

It is clear that any other coordinate system that is just a rotation of coordinates will give the same form because the metric is expressed in scalar products. The curvature constant k is needed to account for a possible, flat, open, or closed universe. The scale factor Q allows for a universe that evolves in time.

This equation yields the Robertson–Walker metric (Robertson, 1935; Walker, 1936). It will soon be seen that the curvature constant k can take on only the values $(1, 0, -1)$. The quasi-translation below takes the origin to $\vec{a}$. It is just an ordinary translation if $k = 0$, and it is obvious that in this case $C_{ij} dx^i dx^j$ and $C_{i'j'} dx^{i'} dx^{j'}$ have the same form. However, this is also true for the other values of k (see Problem 1):

$$x^{i'} = x^i + a^i \left[(1 - kr^2)^{1/2} - (1 - (1 - k(a)^2)^{1/2}) \frac{\vec{a} \cdot \vec{r}}{(a)^2} \right],$$
$$C_{i'j'} dx^{i'} dx^{j'} = \eta_{i'j'} dx^{i'} dx^{j'} + \frac{k\eta_{l'm'} x^{l'} dx^{m'}}{1 - k\eta_{s't'} x^{s'} x^{t'}}. \tag{9.4}$$

The reason constant k can take on only one of the above three values is because r and Q can be renormalized,

$$k > 0, \quad \text{choose } \bar{r} = k^{1/2}r, \quad \bar{Q} = Qk^{-1/2},$$

$$(1 - kr^2)^{-1} = (1 - \bar{r}^2)^{-1},$$

$$(dr)^2 = k^{-1}(d\bar{r})^2,$$

$$d\tau^2 = dt^2 - \bar{Q}^2 \left[\frac{(d\bar{r})^2}{1 - \bar{r}^2} + \bar{r}^2((d\theta)^2 + \sin^2\theta(d\phi)^2) \right],$$

$$k < 0, \quad \text{choose } \bar{r} = |k|^{1/2}r, \quad \bar{Q} = Q|k|^{-1/2},$$

$$(1 - kr^2)^{-1} = (1 + |k|r^2)^{-1} = (1 + \bar{r}^2)^{-1},$$

$$(dr)^2 = |k|^{-1}(d\bar{r})^2,$$

$$d\tau^2 = dt^2 - \bar{Q}^2 \left[\frac{(d\bar{r})^2}{1 + \bar{r}^2} + \bar{r}^2((d\theta)^2 + \sin^2\theta(d\phi)^2) \right].$$

So from now on the bars over r and Q can be dropped.

To get an idea of how the curvature is affected by k, consider the three possible cases. When $k = 0$, the quasi-translation becomes an ordinary translation so that the universe is flat. For other values of k, the metric is diagonal, so the determinant of the metric and the invariant volume element are as follows:

$$(-det(g_{\mu\nu}))^{1/2} = Q^3 \frac{r^2}{(1 - kr^2)^{1/2}} \sin\theta,$$

$$d^4V = Q^3 dt \frac{r^2}{(1 - kr^2)^{1/2}} dr \sin\theta d\theta d\phi.$$

If $k = 1$, choose $\sin u = r$,

$$u = \sin^{-1} r,$$

$$du = (1 - r^2)^{-1/2} dr,$$

$$d^4V = Q^3 dt \sin^2 u du \sin\theta d\theta d\phi.$$

As u increases, it increases faster than $\sin u$. So, for large u, the areas of spheres do not increase as fast as in a flat universe. Such a universe is closed. If $k = -1$, choose $\sinh u = r$,

$$u = \sinh^{-1} r,$$

$$du = (1 + r^2)^{-1/2} dr,$$

$$d^4V = Q^3 dt \sinh^2 u du \sin\theta d\theta d\phi.$$

As u increases, it increases slower than $\sinh u$. So, for large u, the areas of spheres increase faster as compared to a flat universe. Such a universe is open.

The problem of cosmology is to find $Q(t)$ and k. The sections that follow indicate what observations are made and what they lead to. It turns out that $k = 0$, but for historical reasons and to indicate what cosmologists had to go through, it will be written as k for awhile longer.

9.2   The Red Shift

As elements are de-excited they emit characteristic photons, the line spectra of the elements. By comparing light from distant sources with what is observed in the laboratory, it is found that the characteristic wavelengths are always shifted to the red in very distant sources, and the farther away, the more the shift. This doesn't work for close objects like the Andromeda galaxy at a distance of $\approx 2.6 \times 10^6$ ly. Local gravitational effects and other local motion mask the expanding universe effect. This red shift is different from that encountered in Chapter 5, where the light was emitted in stronger or weaker gravity than the observed light.

Assume a light wave crest leaves a typical galaxy at radial coordinate $r = d$ when $t = t_1$ and $Q = Q_1$. The light travels in the $-\hat{e}_r$-direction reaching us when $t = t_0$ and $Q = Q_0$. Many red shifts are observed at t_0, so the red shift is labeled by t_1. The next crest leaves at $t_1 + \delta t_1$, and arrives at $t_0 + \delta t_0$. During the small period of time between crests, Q is essentially constant. As t increases, r decreases and the metric yields for $d\theta = d\phi = 0$,

$$0 = (d\tau)^2 = (dt)^2 - Q^2(dr)^2(1 - kr^2)^{-1},$$

$$dt/Q = -dr(1 - kr^2)^{-1/2},$$

$$\int_{t_1}^{t_0} dt/Q = -\int_d^0 dr(1 - kr^2)^{-1/2} = f(d),$$

$$f[d] = (\sin^{-1} d, d, \sinh^{-1} d), \ k = (1, 0, -1)$$

$$\approx d \quad \text{if } d \ll 1 \text{ or } k = 0, \tag{9.5}$$

$$\int_{t_1+\delta t_1}^{t_0+\delta t_0} dt/Q = f(d) = \int_{t_1}^{t_0} dt/Q \equiv \int_{t_1}^{t_0} dW[t], \tag{9.6}$$

$$W[t_0] - W[t_1] = W[t_0 + \delta t_0] - W[t_1 + \delta t_1],$$

$$W[t_0 + \delta t_0] - W[t_0] = W[t_1 + \delta t_1] - W[t_1],$$

$$\delta t_0 \frac{dW[t]}{dt}\Big|_{t_0} = \delta t_1 \frac{dW[t]}{dt}\Big|_{t_1},$$

$$\frac{\delta t_0}{Q_0} = \frac{\delta t_1}{Q_1}, \quad Q_{0,1} \equiv Q(t_0), Q(t_1).$$

The above equation leads to a simple relation between the frequencies of the emitted and received radiation,

$$\frac{\delta t_0}{\delta t_1} = \frac{\nu_1}{\nu_0} = \frac{\lambda_0}{\lambda_1} = \frac{Q_0}{Q_1} \equiv 1 + z. \tag{9.7}$$

If $z > 0$, then $\lambda_0/\lambda_1 > 1$, which is a red shift. If $z < 0$, then $\lambda_0/\lambda_1 < 1$, which is a blue shift. If for faraway sources there is always a red shift, then $Q_0/Q_1 > 1$ and the universe is expanding. This is what the observations indicate, and z is called the red shift.

In elementary physics classes and indeed when SR was studied, one can show such shifts find a natural interpretation in terms of the Doppler effect. This is not strictly correct as frequencies are affected by the gravitational field of the universe. It is approximately correct, as shown below, for close in sources with small outward radial speeds. From Eq. (2.26), we have

$$\nu_1/\nu_0 = [(1+v)/(1-v)]^{1/2} \approx 1 + v = 1 + z, \quad v = z.$$

The same result is obtained from Eq. (9.7). However, to relate that result to v, the proper distance L_p to the light source is needed. For a light source at radial coordinate d, L_p is given by the metric and Eq. (9.5),

$$L_p = \int_0^d (g_{rr})^{1/2} dr = Q_0 \int_0^d dr(1 - kr^2)^{-1/2} = Q_0 f(d) \tag{9.8}$$

$$\approx Q_0 d \quad \text{if } d \ll 1 \text{ or } k = 0. \tag{9.9}$$

In texts, the proper distance is also called distance now, proper motion distance, co-moving distance, or co-moving radial distance. In this case, the proper velocity of the source is as follows:

$$v_p \equiv \frac{dL_p}{dt} = d\frac{dQ}{dt}\Big|_{t_0} \equiv d\frac{dQ_0}{dt},$$

$$\frac{dQ_0}{dt} = \frac{Q_0 - Q_1}{t_0 - t_1}, \quad t_1 \to t_0,$$

$$\frac{v_p}{d} = \frac{Q_1}{t_0 - t_1}\left[\frac{Q_0}{Q_1} - 1\right]$$

$$= \frac{Q_1}{t_0 - t_1}z \approx \frac{Q_0}{t_0 - t_1}z = \frac{L_p}{t_0 - t_1}\frac{z}{d},$$

$$z \approx v_p.$$

This is the same result as obtained using the Doppler shift. However, it only works for close sources with $L_p/(t_0 - t_1) \approx c = 1$, and the universal expansion can be neglected.

In 1929, E. Hubble observed what he thought were faraway sources, as he had no idea of the size of the universe. We now know they are rather close in objects. He noticed that essentially all the sources showed red-shifted light, indicating an expanding universe. He measured their non-relativistic speeds from the red shift using the Doppler effect and also obtained an estimate of the distances to the sources using Cepheid variable stars, discussed below. The distances were severely overestimated because many of these stars were very dim and part of their light output was scattered by cosmic dust. The observed stars were so close that local motion was hiding the universal expansion. Though his data had large spread, he inferred a linear relation between velocity and distance $z \approx v_p = Hd$. This is the only such relation that would be the same for all observers, independent of position, as illustrated in Weinberg (1977, Fig. 1). These data gave the first value for the Hubble parameter, at times mistakenly called the Hubble constant H. He reasoned that if faraway objects were all rushing away from us, then at

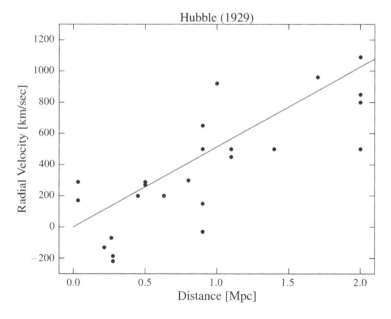

Fig. 9.3   Modern replotting of Hubble's 1929 Fig. 1 by Richard Pogge, The Ohio State University, using the original data of Hubble (1929, Tables 1 and 2).

an earlier time they must have been on top of each other. This allowed him to arrive at an expansion age for the universe $v = d/t_H = Hd$, $t_H = H^{-1}$. The determined age was much lower than the current value, but indicated that it was to be measured in billions of years. This announcement (Hubble, 1929) of a linear relationship between velocity and distance ignited the field. Such measurements still continue with ever improving accuracy (Fig. 9.3).

9.3   Determining Distance

For very close stars, the distance can be determined by parallax. As the earth orbits the sun, the star whose distance is to be measured, appears to move against the background of seemingly fixed distant stars, as indicated in Fig. 9.4. From the extreme wanderings that occur half a year apart, the parallax angle p can be determined from the straight line light paths,

$$b/d_\| = b/d = \tan p \approx p, \quad d_\| = b/p.$$

The origin is at the sun's center, $b = 1.496 \times 10^{11}$ m is the radial coordinate of earth, and d is the radial coordinate of the star. In Newtonian physics, these coordinates are the distances. This formula is good enough because gravity is so weak, and only close stars are considered.

General Relativity (GR) requires taking gravity into consideration. The light paths are geodesics as indicated by the dashed curve in Fig. 9.4. Following the development of Weinberg (1972), light leaves the source at position $\vec{d}$, and eventually reaches us. In the coordinate system $x^{\mu'}$, where the origin is at the light source, the tip of the ray path is at $\vec{r}\,' = \hat{n}r'$. Here $\hat{n}$ is a fixed unit vector and r' is a parameter describing positions along the path. In order to translate to the coordinate system in which the light source is at $\vec{d}$ and the origin is at the center of the sun, the quasi-translation Equation (9.4) must be used, so that the same metric holds for both observers. For this case, use $\vec{a} = \vec{d}$ and $x^i \leftrightarrow x^{i'}$. Thus,

$$\vec{r}(r') = r'\hat{n} + \vec{d}\left[\left(1 - kr'^2\right)^{1/2} - [1 - (1 - kd^2)^{1/2}]\frac{r'\hat{n} \cdot \hat{d}}{d}\right]. \quad (9.10)$$

Only light rays passing close to the origin are observed, so that $\hat{n}$ must nearly point in the $-\hat{d}$-direction. Thus $\hat{n} \approx -\hat{d} + \vec{\delta}$, where $\delta \ll 1$ and $\hat{d} \cdot \vec{\delta} = 0$. In what follows, only first-order terms in δ are kept. In Fig. 9.4, it is the small angle between the light path and $-\hat{d}$ as measured in the $x^{\mu'}$ system.

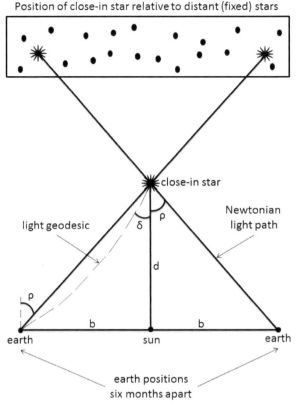

Fig. 9.4   Parallax is the wandering of a close star against the background of fixed, distant stars as the earth orbits the sun. From the extremes of the wanderings, the parallax angle is determined and geometry yields the distance to the star.

If the relation for $\hat{n}$ is inserted into Eq. (9.10), then

$$\vec{r}[r'] = r'(-\hat{d} + \vec{\delta}) + \vec{d}\left[(1 - kr'^2)^{1/2} - [1 - (1 - kd^2)^{1/2}]\frac{-r'}{d}\right]$$

$$= r'\vec{\delta} + \hat{d}\left[-r' + d\left((1 - kr'^2)^{1/2} - [1 - (1 - kd^2)^{1/2}]\frac{-r'}{d}\right)\right]$$

$$= r'\vec{\delta} + \hat{d}[d(1 - kr'^2)^{1/2} - r'(1 - kd^2)^{1/2}]$$

$$\approx d\vec{\delta}, \quad \text{at } r' = d.$$

It is seen that the light ray comes closest to the origin when $r' = d$, as only the very small vector $\vec{\delta}$ is involved.

The impact parameter b is really the proper distance of the light from the origin at this point. Use of the above equation, Eqs. (9.5), (9.6), and (9.8) yield $b = Q_0 d\delta$. Measurements of parallax amount to measurements of the light direction as a function of b. The light ray has a direction, at this point, given by the derivative of the above vector,

$$\left.\frac{d\vec{r}(r')}{dr'}\right|_d = \vec{\delta} + \hat{d}[d[-kr'](1 - kr'^2)^{-1/2} - (1 - kd^2)^{1/2}]|_d$$

$$= \vec{\delta} + \hat{d}(1 - kd^2)^{-1/2}\left[-kd^2 - (1 - kd^2)\right]$$

$$= \vec{\delta} - \hat{d}(1 - kd^2)^{-1/2},$$

$$\vec{e} \equiv -(1 - kd^2)^{1/2} \left.\frac{d\vec{r}(r')}{dr'}\right|_d$$

$$= \hat{d} - (1 - kd^2)^{1/2}\vec{\delta} \approx \hat{e}.$$

To first order in $\vec{\delta}$, the observer's line of sight at this point is in the direction of $\vec{e}$. This direction is opposite the light ray direction at this point. Then the angle between the line of sight and $\hat{d}$ is, noting that the small angle approximation holds,

$$p \approx |\hat{e} - \hat{d}| = (1 - kd^2)^{1/2}\delta = b(1 - kd^2)^{1/2}/(Q_0 d),$$

$$d_{\parallel} \equiv b/p = Q_0 d/(1 - kd^2)^{1/2} \approx Q_0 d = L_p. \tag{9.11}$$

The above is an exact result if $k = 0$. From Euclidean geometry, a source at distance $d_{\parallel}$, with impact parameter $b \ll d_{\parallel}$, has a parallax angle $p = b/d_{\parallel}$. The above expression is general. These distances can only be measured for stars close to us. There $d_{\parallel}$ is the proper distance to the star.

Take $d_{\parallel}$ for the closest stars to be 4 ly, or 3.8×10^{16} m. In this case, $p = 0.83''$. The best satellite telescope Hipparcos (HIPP, 2015) can measure angles to a precision of $0.002''$. So individual stars up to about 1600 ly can be measured by parallax. If you can measure lots of stars N in the same vicinity, for example, a globular cluster or nearby galaxy, the error can be beaten down by a factor $N^{1/2}$. Millions of stars in the Magellanic Clouds have been measured, so that parallax can reach there. The GAIA experiment (Soszynski, 2012) hopes to reach a sensitivity of $20 \times 10^{-6''}$, allowing stars tens of thousands of light years from earth to be measured by parallax.

To go to larger distances, the comparison of absolute luminosity L_{ab} and apparent luminosity flux L_{ap} is exploited, where absolute means the

actual power output, and apparent is the measured power per unit area of
the observing mirror. Dividing by the area standardizes the observations
of different observers. Put the origin at the center of the observing mirror
instead the sun. Let the mirror have radius b, area A, and normal along
the line of sight. The light that reaches the mirror surface lies in a cone, in
the $x^{\mu'}$ system, with half angle δ. The fractional solid angle of this cone is
$\Delta\Omega/4\pi$, where $\Delta\Omega = \pi\delta^2 = \pi[b/(Q_0 d)]^2 = A/(Q_0 d)^2$.

This quantity is the fraction of all isotropically emitted photons that
reach the mirror. It is just the inverse square law. However, due to the
red shift, each photon emitted with frequency $\nu = E/h$, at Q_1 is shifted
to a lower frequency $\nu(Q_1/Q_0) = \nu/(1+z)$, when observed at Q_0. Also,
the photons that were emitted at time intervals δt_1, are received at intervals $\delta t_1 Q_0/Q_1$. Then the absolute luminosity is decreased on traveling the
requisite distance in all directions such that

$$\frac{dE}{dt}\text{(received)} = L_{ab}\left(\frac{Q_1}{Q_0}\right)^2 \frac{A}{4\pi Q_0^2 d^2},$$

$$L_{ap} = \frac{dE}{dt}\text{(received)}/A = \frac{L_{ab}Q_1^2}{4\pi Q_0^4 d^2} \equiv \frac{L_{ab}}{4\pi d_L^2},$$

$$d_L \equiv \left(\frac{L_{ab}}{4\pi L_{ap}}\right)^{1/2} = \frac{Q_0^2 d}{Q_1} \tag{9.12}$$

$$= [Q_0 d]\frac{Q_0}{Q_1} = d_\parallel(1+z) = L_p(1+z). \tag{9.13}$$

The apparent luminosity flux is decreased via the inverse square law,
from which the luminosity distance d_L is defined, as in Euclidean geometry.
The luminosity distance is larger than the proper distance because after the
light is emitted the source and earth separate due to the expanding universe.
The light must travel a longer distance than just the proper distance when
it was emitted, and the light itself is red shifted. For objects with small
red shifts, this doesn't amount to much. However, as we shall see, there are
many objects with large red shifts. For close objects $z \approx 0$ and $d_\parallel = d_L$. So
for sources that allow a parallax distance measurement, L_{ap} leads to the
source's L_{ab}.

There are corrections that experimenters must apply. For example,
detectors are sensitive to only part of the electromagnetic spectrum. Some
light originally leaving the source is red shifted out of the sensitive region,
while other light is red shifted in. Corrections must be made for our rotation
about the Milky Way center and the absorption of light on its journey.

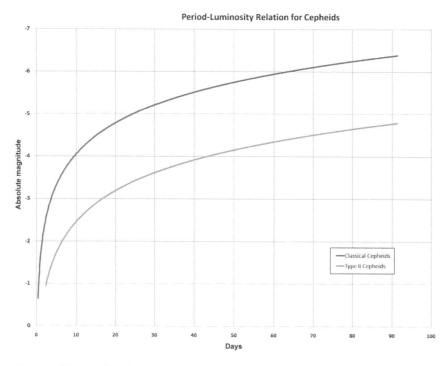

Fig. 9.5 The absolute luminosity–period relation for Cepheid variable stars. The curves indicate the trends of the data. Type II stars are metal rich and brighter than the metal poor type I stars.

Once a large number of stars have their distances measured by parallax, their L_{ab} is determined from L_{ap}. Some stars are variable in their luminosity. For example, Cepheid variables have periods of intensity variation that are closely linked to L_{ab}. The trends in the data are shown in Fig. 9.5. The observation of variable stars with this characteristic, in faraway galaxies, gives their L_{ab} from their period. Measurement of L_{ap} yields d_L for the galaxy. Such observations get astronomers out beyond a number of galaxy clusters. Along the way, the L_{ab} of the brightest stars in galaxies and brightest galaxies in clusters are obtained. Looking out to even fartheraway galaxies and clusters, it can be assumed that their brightest stars and galaxies have the same L_{ab}, and the distance to them is determined by L_{ap}.

Nowadays, type 1 supernovae, that have no hydrogen lines in the visible spectrum, are used as standard candles. Their L_{ab} is calibrated from close in explosions, and used for the most remote sources. They are standard candles because their explosions arise from the same cause. These relatively rare sources were white dwarfs, supported from collapse by electron degeneracy. The latter is a purely quantum effect due to the statistics of identical fermions. They add mass by accretion from nearby objects until carbon fusion is ignited, and the star blows up. It emits as much visible energy as the entire galaxy for a very short time. Their observation gets us out to Bly distances. The WWW has many figures for visualizing the topics in this section.

9.4 Red Shift Versus Distance Relation

Return to the conditions of Section 9.2 where the red shift was defined. An expansion of Q about t_0 can provide a relation for z in terms of the radiation travel time $u_1 \equiv t_0 - t_1$ or the luminosity distance d_L. Assume, the expansion can neglect terms of third order or higher, in the expansion variable $u = t_0 - t$,

$$Q = Q_0 - u \frac{dQ_0}{dt} + \frac{u^2}{2} \frac{d^2 Q_0}{dt^2}, \quad \frac{d^2 Q_0}{dt^2} \equiv \frac{d^2 Q}{dt^2}\Big|_{t_0}$$

$$= Q_0 \left[1 - \frac{u}{Q_0} \frac{dQ_0}{dt} + \frac{u^2}{2Q_0} \frac{d^2 Q_0}{dt^2} \right]$$

$$\equiv Q_0 \left[1 - H_0 u - q_0 H_0^2 u^2 / 2 \right]. \tag{9.14}$$

In the above equation,

$$H = \frac{1}{Q} \frac{dQ}{dt}, \quad -q = \frac{1}{H^2 Q} \frac{d^2 Q}{dt^2} = 1 + \frac{1}{H^2} \frac{dH}{dt}, \tag{9.15}$$

where H is the Hubble parameter with present value H_0 and $-q$ is the acceleration parameter with present value $-q_0$.

Evaluate Eq. (9.14) at time t_1, where $u = u_1$, and find the relation between the travel time and red shift,

$$1 = \frac{Q_0}{Q_1} \left[1 - H_0 u_1 - H_0^2 q_0 u_1^2 / 2 \right]$$

$$= (1 + z) \left[1 - H_0 u_1 - H_0^2 q_0 u_1^2 / 2 \right],$$

$$z = \frac{H_0 u_1 + H_0^2 q_0 u_1^2 / 2}{1 - H_0 u_1 - H_0^2 q_0 u_1^2 / 2}$$

$$\approx \left[H_0 u_1 + H_0^2 q_0 u_1^2 / 2 \right] \left[1 + H_0 u_1 \right]$$

$$\approx H_0 u_1 + H_0^2 u_1^2 (1 + q_0 / 2).$$

This is a quadratic equation that is easily solved for u_1,

$$u_1 = \frac{-H_0 + H_0 \left[1 + 4z(1 + q_0/2)\right]^{1/2}}{2H_0^2(1 + q_0/2)}$$

$$= -H_0 \frac{1 - (1 + 2z(1 + q_0/2)) - [4z(1 + q_0/2)]^2/8)}{2H_0^2(1 + q_0/2)}$$

$$= \frac{z}{H_0}[1 - z(1 + q_0/2)]. \tag{9.16}$$

The relation between d_L and z comes from Eq. (9.6) with $f(d) = d$ and from Eqs. (9.13)–(9.16). Since $du = -dt$,

$$d = \int_{t_1}^{t_0} dt/Q = Q_0^{-1} \int_{t_1}^{t_0} dt \left[1 - H_0 u - H_0^2 q_0 u^2/2\right]^{-1},$$

$$Q_0 d \approx -\int_{u_1}^{0} du \left[1 + H_0 u + q_0 H_0^2 u^2/2 + H_0^2 u^2\right]$$

$$= \int_{0}^{u_1} du \left[1 + H_0 u + (1 + q_0/2)H_0^2 u^2\right] \approx u_1(1 + H_0 u_1/2)$$

$$= \frac{z}{H_0} [1 - z(1 + q_0/2)] [1 + z (1 - z(1 + q_0/2)) /2]$$

$$\approx \frac{z}{H_0} [1 - z(1 + q_0)/2],$$

$$d_L = Q_0 d (1 + z) = \frac{z (1 + z)}{H_0} [1 - z(1 + q_0)/2]$$

$$\approx \frac{z}{H_0} [1 + z(1 - q_0)/2]. \tag{9.17}$$

To lowest order in z, the above is the Hubble relation $z = H_0 d_L$. So the program is to measure z and d_L as accurately as possible, for many objects, to determine H_0 and q_0. To measure the latter you need to measure at very large distances where $z > 0.5$. Here, type 1 supernovae serve as standard candles. A number of them, at various d_L, are needed to determine the shape of the curve. For H_0, objects with $0.05 \leq z \leq 0.1$ are needed, to confirm the approximate linearity of the d_L versus z relation. For smaller red shifts, you could be measuring a local velocity rather than the universal expansion. Even for the Virgo cluster, that was beyond Hubble's reach, $d_L \approx 16.5$ Mpc and $z \approx 0.04$.

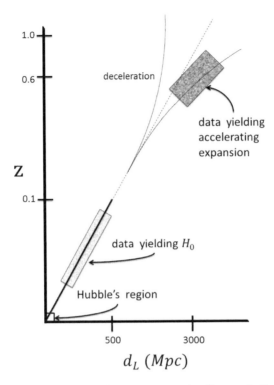

Fig. 9.6   Supernova data for red shift versus luminosity distance indicate that the universe is undergoing an accelerating expansion (see text).

A sample of the modern data is schematically shown in Fig. 9.6. Hubble's data region is a very small area near the origin. The two groups providing the data are the HighZ Supernova Search Team (Riess, 1998), and the Supernova Cosmology Project (Perlmutter, 1999). The data are shown schematically as rectangular bands, and especially at the highest red shift values, indicate that a straight line fit fails. An accelerating expansion is a better explanation. The data are fit to Robertson–Walker dynamic models for the matter and dark energy densities as discussed below in Section 9.6. The analysis from the PLANCK Collaboration (2015) yields the following:

$$H_0 = (67.8 \pm 1.7\%) \ \text{km s}^{-1} \ \text{Mpc}^{-1}, \tag{9.18}$$

$$t_{H_0} = H_0^{-1} = (14.4 \pm 1.7\%) \times 10^9 \ \text{y}, \tag{9.19}$$

$$q_0 \approx -0.54. \tag{9.20}$$

The result for q_0 is based on a calculation using the best fits for the matter and dark energy densities (see Problem 12). However, just fitting would have found $-q_0 > 0$. Thus $\frac{d^2 Q_0}{dt^2} > 0$. The expansion is accelerating. The universe is open and will expand forever. As shall be seen, a way to explain this in GR is to keep a nonzero cosmological constant Λ in Eq. (5.9). The contribution from the cosmological constant is a form of non-understood "dark" energy that provides a negative pressure. In a younger universe, dark energy was not so important, however, it has now taken over.

This leads to a note worthy item. The proper velocity $v_p = d\frac{dQ}{dt}$ is going to get larger and larger as time increases. A future observer may see $v_p > 1$. This doesn't conflict with relativity. The photons emitted by an object still have $c = 1$, however, space is being created between objects. This is absent from relativity theory, and that's what's making $v_p > 1$.

9.5 Fluids

In order to solve Einstein's equation (5.9), the energy–momentum tensor $T_{\mu\nu}$ is needed. Also, even though gravity is weak, GR is required. For example, if the universe had a uniform mass density $\bar\rho$, the quantity $M/r = 4\pi\bar\rho r^2/3 > 1$, at some r. In order to make headway, the many bodies making up the universe are subject to a simplifying assumption, namely that they constitute a perfect fluid. Then an momentum–energy tensor, that makes testable predictions, can be obtained.

In the SR frame O where all the particles are at rest the fluid is called dust. In this frame, the number density of dust particles is $N/V = n\,\mathrm{m}^{-3}$. In another SR frame O$'$, where the particles are moving with velocity $\vec{v}\,'$, the volume element is contracted by a factor $1/\gamma = (1 - |\vec{v}'|^2)^{1/2}$. In O$'$ $n' = \gamma n$.

The flux of particles across a surface is the number crossing the surface in the direction of the normal to the surface per unit area per unit time. Thus, all particles within a distance $v^{\bar{i}'}\Delta t'$ of the surface, where $v^{\bar{i}'}$ is the speed in the normal direction and within area ΔA that defines the size of the surface, will cross the surface in time period $\Delta t'$,

$$f = (n'v^{\bar{i}'}\Delta t'\Delta A)/(\Delta t'\Delta A)$$
$$= \gamma n v^{\bar{i}'}.$$

Note that the vector $N^{\bar{\mu}'} = nU^{\bar{\mu}'}$ combines both the flux and number density,

$$N^{\bar{0}'} = nU^{\bar{0}'} = \gamma n,$$

$$N^{\bar{i}'} = nU^{\bar{i}'} = \gamma n v^{\bar{i}'},$$

$$N^{\bar{\mu}'} N_{\bar{\mu}'} = \gamma^2 n^2 (-1 + (v')^2) = -n^2. \tag{9.21}$$

This result shows that n is an invariant, just as a particle's momentum vector $P^\mu = mU^\mu$, yields $P^\mu P_\mu = -m^2$ for the particle's invariant rest mass.

In frame O, the energy of each particle is just the rest energy. If all the particles had the same rest mass m, the total energy density would be $\rho = nm$. In the frame O', the energy would be m increased by a factor γ. The number density would be similarly increased, thus $\rho' = \gamma^2 \rho$. The energy–momentum tensor is the mathematical way to express these things,

$$T^{\bar{\mu}' \bar{\nu}'} = P^{\bar{\mu}'} N^{\bar{\nu}'} = \rho U^{\bar{\mu}'} U^{\bar{\nu}'}. \tag{9.22}$$

It's obvious that this tensor is symmetric and its divergence is zero. Physically, it represents the flux of $P^{\bar{\mu}'}$ across a surface of constant $x^{\bar{\nu}'}$.

For example, dust has only one nonzero element $T^{\bar{0}\bar{0}} = \rho$. Using Eqs. (1.9)–(1.11) and (2.10), the tensor in O' is as follows:

$$T^{\bar{\mu}' \bar{\nu}'} = x^{\bar{\mu}'}{}_{,\bar{0}}\, x^{\bar{\nu}'}{}_{,\bar{0}}\, \rho,$$

$$T^{\bar{0}' \bar{0}'} = x^{\bar{0}'}{}_{,\bar{0}}\, x^{\bar{0}'}{}_{,\bar{0}}\, \rho = \gamma^2 \rho,$$

$$T^{\bar{0}' \bar{i}'} = x^{\bar{0}'}{}_{,\bar{0}}\, x^{\bar{i}'}{}_{,\bar{0}}\, \rho = \gamma^2 \rho v^{\bar{i}'},$$

$$T^{\bar{i}' \bar{j}'} = x^{\bar{i}'}{}_{,\bar{0}}\, x^{\bar{j}'}{}_{,\bar{0}}\, \rho = \gamma^2 \rho v^{\bar{i}'} v^{\bar{j}'}.$$

The element $T^{\bar{i}' \bar{j}'}$ being a momentum density change across a surface is proportional to the force on the surface. In a liquid, this is equivalent to the pressure $p = \text{Energy/Volume} = \text{Force/Area}$.

To get to the properties of a perfect fluid, consider a general fluid and its thermodynamic properties. In this case, one can consider the rest frame of each fluid element. This is the element's momentarily co-moving reference frame (MCRF). Since fluids can be accelerated, this frame may not remain the MCRF at all times. Other elements have different MCRF's. In this frame, there is no bulk flow and no spatial momentum in the particles. All scalar quantities associated with a fluid element are defined to be the values in the MCRF. These include: rest energy Nm, rest energy density mn, total energy density ρ, number density n, internal energy per particle $\rho/n - m$, temperature T, pressure p, and entropy per particle S. In addition, there are the vectors $U^{\bar{\mu}}$ and $N^{\bar{\mu}}$.

In the element's MCRF energy E can be exchanged by the fluid absorbing or emitting heat $\pm dQ$, and by doing work or by having work done on it $\pm dW = \pm p dV$. When this is allowed, other elements of $T^{\mu\nu}$ become finite. For example, the first law of thermodynamics yields, for an element with N particles,

$$
\begin{aligned}
dE &= dQ - p dV, \\
V &= N/n, \quad dV = -N dn/n^2, \\
E &= \rho V, \quad dE = V d\rho + \rho dV, \\
dQ &= V d\rho + \rho dV + p dV \\
&= \frac{N}{n}\left(d\rho - (\rho+p)\frac{dn}{n}\right), \\
dq &\equiv dQ/N, \\
n dq &= d\rho - (\rho+p)\frac{dn}{n} = nT dS,
\end{aligned}
\tag{9.23}
$$

where T is the temperature in K(elvin) and S is the entropy per particle.

For the energy–momentum tensor this means $T^{\bar{0}\bar{0}} = \rho$, the energy density. Since none of the particles in the element have spatial momentum, the energy flux $T^{\bar{0}\bar{i}}$ is a heat conduction term as is $T^{\bar{i}\bar{0}}$. In the case of a perfect fluid, there is no heat conduction in the MCRF, so that $T^{\bar{0}\bar{i}} = T^{\bar{i}\bar{0}} = 0$. Viscosity is a force parallel to the interface between elements. Its absence, in a perfect fluid, means that the force is perpendicular to the interface and $T^{\bar{i}\bar{j}} = 0$, $\bar{i} \neq \bar{j}$. This makes the energy–momentum tensor for a perfect fluid diagonal. Also there is no preferred direction so all $T^{\bar{i}\bar{i}}$ have the same value. This feature will be preserved for the fluid as a whole as it is true for each element,

$$
\begin{aligned}
T^{\bar{0}\bar{0}} &= \rho, \\
T^{\bar{i}\bar{i}} &= p, \quad \text{thus,} \\
T^{\bar{\mu}\bar{\nu}} &= (\rho + p)U^{\bar{\mu}}U^{\bar{\nu}} + p\eta^{\bar{\mu}\bar{\nu}}, \quad U^{\bar{i}} = 0.
\end{aligned}
\tag{9.24}
$$

In the above equation, $U^{\bar{i}} \to 0$ because no direction is preferred. The above equation is a tensor equation and is the equation in a locally inertial frame.

For GR in general,

$$T^{\mu\nu} = (\rho + p)U^\mu U^\nu + pg^{\mu\nu}, \quad U^i = 0. \tag{9.25}$$

As previously discussed, this tensor and the vector that forms it have zero divergence,

$$T^{\nu\mu}{}_{;\mu} = 0 \tag{9.26}$$

$$U^\mu{}_{;\mu} = N^\mu{}_{;\mu} = 0. \tag{9.27}$$

9.6 Robertson–Walker Einstein Dynamics

The nonzero C symbols, Ricci tensor and scalar, and Einstein tensor for the Robertson–Walker metric were evaluated in Problems 4.7 and 4.8. The following results were obtained for the C symbols,

$$\Gamma^0_{ii} = g_{ii}\frac{1}{Q}\frac{dQ}{dt}, \quad \Gamma^i_{0i} = \frac{1}{Q}\frac{dQ}{dt},$$

$$\Gamma^r_{rr} = \frac{kr}{1-kr^2}, \quad \sin^2\theta\Gamma^r_{\theta\theta} = -r\left(1-kr^2\right)\sin^2\theta = \Gamma^r_{\phi\phi}, \tag{9.28}$$

$$\Gamma^\theta_{\theta r} = r^{-1}, \quad \Gamma^\theta_{\phi\phi} = -\sin\theta\cos\theta, \quad \Gamma^\phi_{\phi r} = r^{-1}, \quad \Gamma^\phi_{\phi\theta} = \cot\theta.$$

These allowed the calculation of the Ricci tensor and scalar,

$$R_{00} = = -3\frac{1}{Q}\frac{d^2Q}{dt^2} = R^{00},$$

$$R_{ii} = g_{ii}\frac{1}{Q^2}\left[2\left(k + \left(\frac{dQ}{dt}\right)^2\right) + Q\frac{d^2Q}{dt^2}\right] = g_{ii}^2 R^{ii} = g_{ii}R^{ii}/g^{ii}, \tag{9.29}$$

$$R = g^{\mu\nu}R_{\mu\nu} = g^{\mu\mu}R_{\mu\mu} = 6\frac{1}{Q^2}\left[Q\frac{d^2Q}{dt^2} + \left(\frac{dQ}{dt}\right)^2 + k\right],$$

and the Einstein tensor elements,

$$G_{\mu\nu} = R_{\mu\nu} - g_{\mu\nu}R/2, \quad G^{\mu\nu} = R^{\mu\nu} - g^{\mu\nu}R/2,$$

$$G_{00} = 3\frac{1}{Q^2}\left[\left(\frac{dQ}{dt}\right)^2 + k\right], \tag{9.30}$$

$$G_{ii} = -g_{ii}\frac{1}{Q^2}\left[2Q\frac{d^2Q}{dt^2} + \left(\frac{dQ}{dt}\right)^2 + k\right].$$

The Einstein equation for the 00 element is as follows:

$$G^{00} + \Lambda g^{00} = 8\pi T^{00} = 8\pi \left[(\rho + p)U^0 U^0 + p g^{00}\right],$$

$$G^{00} = 8\pi \left[(\rho + p)U^0 U^0 + p g^{00} - \frac{\Lambda}{8\pi} g^{00}\right].$$

Writing the cosmological constant term in the form of a perfect fluid is revealing,

$$G^{00} \equiv 8\pi \left[(\rho + p)U^0 U^0 + p g^{00} + (\rho_\Lambda + p_\Lambda) U^0 U^0 + p_\Lambda g^{00}\right],$$

$$\rho_\Lambda + p_\Lambda = 0, \quad p_\Lambda = -\frac{\Lambda}{8\pi} = -\rho_\Lambda, \tag{9.31}$$

$$G^{00} = 8\pi \left[\rho + \rho_\Lambda\right].$$

In these equations, ρ is the energy density due to matter and radiation, so that $\rho = \rho_R + \rho_M$, and p is the ordinary pressure associated with these energies. They can only be functions of time. However, positive Λ yields a positive "dark energy density" and a negative "dark pressure." These stay constant as the universe expands and because the universe is now so large, Λ drives the expansion. It is interesting to recall that in Problem 5.4 the Schwarzschild problem with Λ included was solved. In weak gravity Λ led to a repulsive Newtonian force.

Using Eq. (9.15), $g^{00} = -1$, $\left(U^0\right)^2 = 1$, and inserting the explicit form for G^{00} into Eq. (9.31) yields

$$3\frac{1}{Q^2}\left[\left(\frac{dQ}{dt}\right)^2 + k\right] = 8\pi \left[\rho + \rho_\Lambda\right],$$

$$\left(\frac{dQ}{dt}\right)^2 = -k + 8\pi Q^2 \left[\rho + \rho_\Lambda\right]/3, \quad \text{or,} \tag{9.32}$$

$$\frac{3H^2}{8\pi} = -\frac{3k}{8\pi Q^2} + \left[\rho + \rho_\Lambda\right]. \tag{9.33}$$

The present-day critical energy density ρ_c can be defined,

$$\rho_c = \frac{3H_0^2}{8\pi}, \quad \rho_c(\text{MKS}) = \frac{3H_0^2 c^2}{8\pi G} = 7.66 \times 10^{-10} \text{ Jm}^{-3}. \tag{9.34}$$

At the present time, Eq. (9.33) can be rewritten,

$$1 = \frac{1}{\rho_c}\left[-\frac{3k}{8\pi Q_0^2} + \rho_0 + \rho_\Lambda\right] \equiv \Omega_k + \Omega_{R0} + \Omega_{M0} + \Omega_\Lambda. \tag{9.35}$$

The data of the PLANCK collaboration (2015) yield

$$\Omega_k = 0, \quad \Omega_{R0} \approx 5.5 \times 10^{-5}, \quad \Omega_{M0} = 0.308, \quad \Omega_\Lambda = 0.692. \qquad (9.36)$$

These results are best fits to lots of data including red shift versus distance and the very small measurable non-uniformity of the CMB. However, the observed density ratio for the ordinary luminous matter is $\Omega_{OM0} \approx 0.05$. So there must be a form of "dark" matter, that only interacts with ordinary matter via gravitation, such that $\Omega_{DM0} \approx 0.253$. It's an interesting universe; the stuff that the stars and ourselves are made of accounts for only five percent of its energy density. The rest is now made up of non-understood dark matter and energy.

Experimenters in deep underground laboratories have mounted experiments to directly detect the rare interactions of dark and ordinary matter. In space, they look for gamma rays from dark matter, anti-matter annihilation, but so far there is no positive result. There is good indirect evidence for dark matter. For example, much stronger than expected gravitational lensing by galaxies of distant objects is observed. Another example concerns stars far from the center of galaxies close to us. As the galaxies are close, red shifts due to the universal expansion are extremely small. The stars far from the galaxy center are rotating about the galaxy center of mass, located near the luminous center of the galaxy. The velocity of these stars can be measured by observing Doppler-shifted spectral lines of the elements. Some stars will have blue-shifted lines, while stars on the other side of the center will have red-shifted lines.

In order to keep things simple, suppose the bulk of the galactic luminous mass M is spherically distributed and the stars at distance R from the center are well outside of this mass distribution. Let the velocity relative to the galactic center be V. Then from elementary Newtonian considerations, the centripetal acceleration is due to gravity,

$$V^2/R = M/R^2, \quad V \propto R^{-1/2}.$$

However, the observations on Andromeda (Rubin, 1970), are in conflict with this expectation. If only the luminous matter was available, the stars should be flying off from the galaxy. They found that the velocity varied from being flat to $V \propto R$. The latter behavior would be expected for stars moving inside an unseen uniform mass density ρ,

$$V^2/R = \left[(4\pi/3)\,\rho R^3\right]/R^2 = (4\pi/3)\,\rho R,$$
$$V^2 = (4\pi/3)\,\rho R^2, \quad V \propto R.$$

This enormous sea of unseen matter in which the galaxy is contained is termed "dark" matter.

You might ask how the luminous mass of stars is determined? Many stars exist as binary pairs. By determining their orbits about the common center of mass, the masses may be obtained. The mass, luminosity, and temperature of the stars form a profile of these stars. It is found that all stars follow a sort of universal distribution on a plot of luminosity versus temperature. So the masses of stars that are not part of a binary pair are deduced from their position on the plot, called a Hertzsprung-Russell diagram (see Kaufman, 1985).

Return to the Einstein field equations. Since Q gives the size of the universe, the quantity $Q^2 \rho_{OM0+DM0} \to Q^2 M/Q^3 = M/Q$ decreases as Q increases. However, $Q^2 \rho_\Lambda$ increases as Q^2 increases because ρ_Λ is constant. This term has taken over, as the dominant energy density as Q has increased so much. The radiation energy density is small now because the universe has expanded and cooled, but as seen below was once the dominant term. From Eqs. (9.25) and (9.30)–(9.32), the Einstein equation for G^{rr} is as follows:

$$G^{rr} + 8\pi \frac{\Lambda}{8\pi} g^{rr} = 8\pi T^{rr} = 8\pi p g^{rr}.$$

Inserting the explicit form for G^{rr} yields

$$0 = g^{rr} \left[-\frac{1}{Q^2} \left(k + \left(\frac{dQ}{dt} \right)^2 + 2Q \frac{d^2 Q}{dt^2} \right) + 8\pi \left(\frac{\Lambda}{8\pi} - p \right) \right]$$

$$= g^{rr} \left[-\frac{8\pi}{3} \left(\rho + \frac{\Lambda}{8\pi} \right) - \frac{2}{Q} \frac{d^2 Q}{dt^2} + 8\pi \left(\frac{\Lambda}{8\pi} - p \right) \right]$$

$$= -\frac{8\pi}{3} \left(\rho + \frac{\Lambda}{8\pi} \right) - \frac{2}{Q} \frac{d^2 Q}{dt^2} + 8\pi \left(\frac{\Lambda}{8\pi} - p \right),$$

$$\frac{2}{Q} \frac{d^2 Q}{dt^2} = 8\pi \left[-(\rho/3 + p) + (2/3) \frac{\Lambda}{8\pi} \right] > 0.$$

The second derivative is positive now because $p \approx 0$ and dark energy dominates. Thus, the expansion will continue at an ever faster rate. If there was no dark energy content, a positive $\frac{d^2 Q}{dt^2}$ would not be observed, and the universe would eventually collapse with a big crunch.

9.7   The Early Universe

In the very young smaller universe, the energy density was dominated by the radiation. At present, the universal CMB presents the most perfect black-body spectrum or Planck distribution. Penzias and Wilson, radio astronomers at Bell Labs, discovered this spectrum in 1964–65. The interesting story of that discovery is described by S. Weinberg (1977). In the Planck distribution, the energy per unit volume in the range of wavelengths between λ and $\lambda + d\lambda$ is given by

$$du = 16\pi^2 \hbar c \frac{d\lambda}{\lambda^5} \left(\exp \left[\frac{2\pi \hbar c}{\lambda k_B T_0} \right] - 1 \right)^{-1},$$

$$u = \int_0^\infty du = aT^4 \tag{9.37}$$

$$= \frac{8\pi^5 (k_B T)^4}{15 (2\pi \hbar c)^3} = 7.565 \times 10^{-16} T^4 \text{ Jm}^{-3}.$$

The above formula is written in MKS units so it appears as you have learned it in elementary thermodynamics. Problem 8 explores the conversion to natural units. The temperature must be given in Kelvin. At present, $T_0 = 2.726$ K. The distribution vanishes at both $\lambda = 0, \infty$, and reaches a maximum at $\lambda \approx 1.263 \hbar c/(k_B T)$.

In the past, the size of the universe was smaller by the cube of the scale ratio $f = Q/Q_0$. Also the wavelength of each photon would be smaller by a factor f because as time decreases, the space between typical particles decreases. The photons' wavelengths are compressed by the same factor. This is just the same as saying the photons have been red shifted to the values we now measure. As a photon's energy goes inversely as its wavelength, the energy density is larger by the factor f^{-4}. In the past,

$$du' = \frac{du}{f^4} = 16\pi^2 \hbar c \frac{d\lambda}{f^4 \lambda^5} \left(\exp \left[\frac{2\pi \hbar c}{\lambda k_B T_0} \right] - 1 \right)^{-1},$$

$$\lambda' \equiv f\lambda, \quad d\lambda' = f d\lambda,$$

$$du' = 16\pi^2 \hbar c \frac{d\lambda'/f}{\lambda'^5/f} \left(\exp \left[\frac{2\pi \hbar c f}{\lambda' k_B T_0} \right] - 1 \right)^{-1}$$

$$= 16\pi^2 \hbar c \frac{d\lambda'}{\lambda'^5} \left(\exp \left[\frac{2\pi \hbar c}{\lambda' k_B T'} \right] - 1 \right)^{-1}.$$

So in the early universe the Planck distribution also described the radiation, but with a higher temperature $T' = T_0/f$. As the radiation energy density is directly propositional to T^4, in that epoch the radiation dominates, and the Einstein equation for G^{00} yields

$$G^{00} = 3 \left(\frac{1}{Q} \frac{dQ}{dt} \right)^2 = 8\pi \rho_R,$$

$$\left(\frac{dQ}{dt} \right)^2 = H_0^2 Q^2 \frac{\rho_R}{\rho_c}$$

$$= H_0^2 Q^2 \frac{\rho_{R0}}{\rho_c} \left(\frac{Q_0}{Q} \right)^4$$

$$= H_0^2 Q_0^4 \Omega_{R0} Q^{-2} \equiv b^2 Q^{-2},$$

$$\frac{dQ}{dt} = b\frac{1}{Q}, \quad \text{so } Q dQ = b dt,$$

$$Q^2/2 = b(t - t'). \tag{9.38}$$

So $Q = 0$ was achieved at some finite time in the past. Taking $t' = 0$, leads to $Q = 0$ when $t = 0$ and $Q \to t^{1/2}$. So there had to be a start to the expansion. The start is called the "big bang," and time started at that moment.

The very early universe, say way before 0.01 s, is not as complicated as now. Then there were just elementary particles and anti-particles in thermal equilibrium with the photons, at a very high temperature. There were equal numbers of all types of particles as the energetic photon–photon collisions could produce all types of particle–anti-particle pairs. The pairs rapidly annihilated into photons. Since expansion is a cooling process, the density was also decreasing. When T became so low that even electron–positron pairs could no longer be created in photon–photon collisions, matter particles and anti-particles went through their final annihilation. Problem 2.15 showed that the photon energy was then 0.511 MeV, the temperature was 0.59×10^{10} K, and $Q/Q_0 = 4.6 \times 10^{-10}$. As we are here and there is no observed evidence for primordial anti-matter, some matter was left over. Using the energy density data given in the previous section and the most simplistic model for the nucleon and photon energies, a nucleon to CMB photon fraction of $\approx 0.23 \times 10^{-9}$ is obtained in Problem 9. More refined models obtain a fraction of $\approx 0.61 \times 10^{-9}$. Thus, for every order 10^9 anti-matter particles, there must have been order $10^9 + 1$, matter particles. No one knows why!

All the free matter particles except the protons and electrons are unstable. For example, except for the neutron, the muon has the longest mean life 660 m or 2.2 μs. So all of these exotic particles are gone well before the universe age is 0.01 s. They are now produced at accelerators or by astrophysical processes. The free neutron has a mean life of 880 s, so many of them stick around much longer, but some decay into protons $n \to pe\bar{\nu}$. The number of protons increases and that of neutrons decreases. The universe is still too hot for nuclei more complicated than ^1H to form, and not be blasted apart by the radiation. However, at a universe age of between 180 s and 240 s, the nucleons could freeze into stable complex nuclei. Calculations predicting 76% ^{1}H and 24% ^{4}He are in agreement with observation.

There are only minute amounts of other hydrogen, helium, and lithium isotopes because their binding energies/nucleon are so much smaller than that of ^{4}He. Essentially all the neutrons froze into ^{4}He. You may ask why weren't there heavier nuclei, with even larger binding energies per nucleon formed? Well, in her wisdom, Mother Nature did not make any stable nuclei with five or eight nucleons. So it wasn't possible for ^{4}He to pick up another nucleon, nor was it possible for two ^{4}He nuclei to combine. Later on very massive stars formed and died, the heavier elements being created in fusion reactions in the stars, reactions between stars, and during supernova. The latter causes the star's contents to be spewed out into space, and become the material of a later generation of stars — the sun.

It is still too hot for atoms to form because their binding energies are in the eV range while nuclear binding energies are in the MeV range. So the photons continue scattering from the electrons. At ≈ 0.37 My, atoms formed, and the universe became transparent to the photons. As the universe expanded, their wavelengths stretched and their temperature decreased, while maintaining their Planck distribution. This is the present-day CMB. The universe became transparent to neutrinos before photons as their probability for interaction is smaller. So the former have a lower temperature. However, such times are so small that they don't effect these calculations. It would be nice to observe the neutrino black-body spectrum, and measure a lower T than we do for the photons, but that seems an impossibility at present.

In this early time era, Eq. (9.38) yields

$$Q = at^{1/2}, \quad H = \frac{1}{Q}\frac{dQ}{dt} = \frac{1}{2t}.$$

As you go back in time, H is increasing rapidly. Also if $t = 0$ is the beginning, then radiation that left then, that is observed at time t, defines the horizon distance $D_H(t)$. This is the proper distance to the radiation source that emitted photons at $t = 0$, that are observed at t. For that radiation,

$$(d\tau)^2 = (dt)^2 - Q^2(t)(dr)^2 = 0,$$

$$dr = \frac{dt}{Q}, \quad \int_0^{r_{\max}(t)} dr = \int_0^t \frac{dt'}{Q'},$$

$$D_H(t) = Q \int_0^{r_{\max}(t)} dr = Q \int_0^t \frac{dt'}{Q'} \qquad (9.39)$$

$$= t^{1/2} \int_0^t dt' t'^{-1/2} = 2t.$$

So as you go back to very small times, the size is decreasing like $t^{1/2}$, but the horizon is decreasing like t. This means the horizon is getting smaller faster than the size of the universe. So less of the universe could be observed. For example, for $t = (0.0001, 0.01)\,\mathrm{s}$, $Q/[aD_H] = 0.5t^{-1/2} = 0.5(100, 10)\,\mathrm{s}^{-1/2}$.

If there was no causal connection between particles at the beginning, how can one account for the flatness and the observation that the background radiation is the same in all directions? This is especially true for opposite directions. A model called "Inflation" (Guth, 1998) postulates that before the Planck time $t_P = \left(\hbar G/c^5\right)^{1/2} \approx 0.5 \times 10^{-43}\,\mathrm{s}$ the universe was causally connected and went through a vast, extremely rapid inflation in size before it settled down to the expansion path outlined in this chapter. The Planck time is constructed from the three basic constants, G, $\hbar$, and c. A Planck length $L_P = ct_P \approx 2 \times 10^{-35}\,\mathrm{m}$ and a Planck mass, $M_P = (\hbar c/G)^{1/2} \approx 1 \times 10^{22}\,\mathrm{MeV}/c^2$ may similarly be constructed. Since $\hbar$ is present, a quantum theory of gravity is required to understand the beginning.

9.8   Matter and Dark Energy Domination

For most of the history of the universe, the dominant energies have been a combination of matter and dark energy Λ. Dark energy is now taking over. However, the period where matter was dominant and dark energy was neglected can be examined,

$$\left(\frac{dQ}{dt}\right)^2 = H_0^2 Q^2 \frac{\rho_{OM+DM}}{\rho_c} \equiv H_0^2 Q^2 \frac{\rho_M}{\rho_c}$$

$$= H_0^2 Q^2 \frac{\rho_{M0}}{\rho_c} \left(\frac{Q_0}{Q}\right)^3$$

$$= H_0^2 Q_0^3 \Omega_{M0} Q^{-1} \equiv b^2 Q^{-1},$$

$$Q^{1/2} dQ = b\, dt, \ 2Q^{3/2}/3 - b\,(t - t''), \ t'' \approx 0,$$

$$Q \propto t^{2/3}, \ \frac{dQ}{dt} \propto 2t^{-1/3}/3,$$

$$H = \frac{1}{Q}\frac{dQ}{dt} = \frac{2}{3t}, \ t = \frac{2}{3}H^{-1}.$$

Here, $t'' \approx 0$, compared to the billions of years that the radiation was no longer important. In this case, $t_{H_0} = (2/3)\,H_0^{-1} = 9.7$ By. The factor $2/3$ due to GR was not included in Hubble's evaluation of the age of the universe. This was the expectation before dark energy was added to the energy–momentum tensor. The calculation above really jammed up the works for awhile because some of the globular clusters in our galaxy are measured to be older. It takes the cosmological constant or dark energy to make an older universe. If dark energy becomes completely dominant, then you get $H(\text{future}) = \text{constant}$. So the critical density will not change in the future.

This leads to a perplexing problem. One assumes that the dark energy has something to do with the vacuum. However, if one worked out vacuum energy, that involves all those virtual particles; it would be seen that the measured dark energy is smaller than the calculated value, by more than 70 orders of magnitude.

If all energy densities are kept and a change of variable $u \equiv Q/Q_0$, $dQ/Q = du/u$ employed,

$$\left(\frac{1}{Q}\frac{dQ}{dt}\right)^2 = H_0^2 \left(\rho_R + \rho_M + \rho_\Lambda\right)/\rho_c$$

$$= H_0^2 (\Omega_\Lambda + \Omega_{M0}\,(Q_0/Q)^3 + \Omega_{R0}\,(Q_0/Q)^4)$$

$$= H_0^2 \left(\Omega_\Lambda + \Omega_{M0}u^{-3} + \Omega_{R0}u^{-4}\right),$$

$$\frac{1}{Q}\frac{dQ}{dt} = H = H_0 \left(\Omega_\Lambda + \Omega_{M0}u^{-3} + \Omega_{R0}u^{-4}\right)^{1/2}, \qquad (9.40)$$

$$dt = \frac{du}{H_0 u \left(\Omega_\Lambda + \Omega_{M0} u^{-3} + \Omega_{R0} u^{-4} \right)^{1/2}}. \qquad (9.41)$$

Initially $u = 0$ while at the present time $u = 1$. Thus,

$$t_0 = H_0^{-1} \int_0^1 \frac{du}{u} \left(\Omega_\Lambda + \Omega_{M0} u^{-3} + \Omega_{R0} u^{-4} \right)^{-1/2} = 13.8 \text{ By},$$

when the integral is performed numerically. This indicates the expansion time is close to the Hubble expansion time H_0^{-1}. One can differentiate Eq. (9.40) with respect to time and calculate the present value for q_0 (see Problem 12).

The proper distance to a source that emitted light at $t \approx 0$, that is just received today, is the distance to the horizon. From Eqs. (9.39) and (9.41), this is

$$D_H (t) = Q \int \frac{dt'}{Q'} = \frac{Q}{Q_0} \int dt' \left(\frac{Q'}{Q_0} \right)^{-1},$$

$$D_{H0} = H_0^{-1} \int_0^1 \frac{du}{u^2} \left(\Omega_\Lambda + \Omega_{M0} u^{-3} + \Omega_{R0} u^{-4} \right)^{-1/2} = 46 \text{ Bly}, \qquad (9.42)$$

when the integral is done numerically. However, there are objects farther away, so that their light will be viewed in the future. There are also objects that will never be viewed. This is because the speed of light is unity, but the space between objects is expanding more and more rapidly. No observer will see the light from such separated objects. We can't tell how large the universe is at any time. This partially explains why the night sky is dark, as every line of sight does not end on a star. However, a better answer lies in the previously discussed scenario: the amount of matter that gravity can clump into stars is about a billion times smaller than what was available in the early universe. The matter available for stars is now so dilute that every line of sight would not end on a star. If it did, the universe would be too hot for life as we know it.

One can see what the past was and what the future holds by changing the integration upper limit. Figure 9.7 shows that with increasing time, H has decreased rapidly, turned, and is entering its ultimate fate of constancy. When the universe is twice as large as now $H \approx 0.85 H_0$. Figure 9.8 shows that at that size the age will be $\approx 1.8 t_0$. When the universe was ten times smaller than it is now, its age was $\approx 0.25 t_0$, but when the universe is ten times as large as now, the age will be $\approx 3.6 t_0$. This is a potent illustration

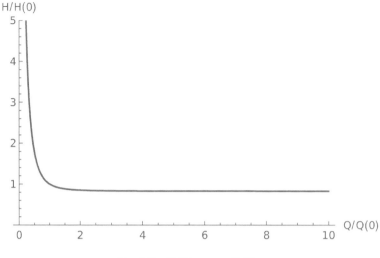

Fig. 9.7 H/H_0 versus Q/Q_0.

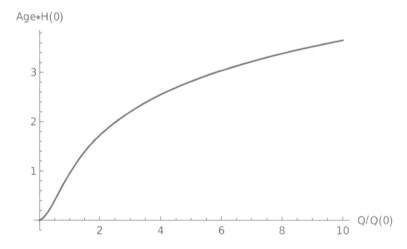

Fig. 9.8   Age $* H_0$ versus Q/Q_0, where Age is the expansion age.

of the accelerating expansion. The size is increasing faster than the aging. By the time that size is reached $H \approx 0.8 H_0$, and will remain constant; the horizon distance will be $D_H \approx 13 D_{H0}$, and increasing linearly with age. It's intriguing to imagine the view (Fig. 9.9).

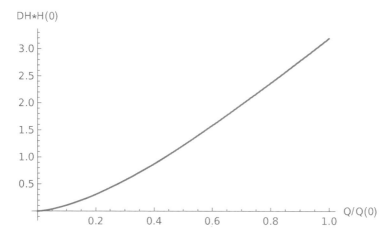

Fig. 9.9 $D_H * H_0$ versus Q/Q_0, where D_H is the horizon distance.

Problems

1. Consider the quasi-translation of coordinates to new origin at $\vec{a}$,

$$\vec{r}\,' = \vec{r} + \vec{a}\left[(1 - kr^2)^{1/2} - [1 - (1 - ka^2)^{1/2}]\frac{\vec{r}\cdot\vec{a}}{a^2}\right],$$

where $r^2 = \vec{r}\cdot\vec{r}$ and $a^2 = \vec{a}\cdot\vec{a}$. Show that this relation leads to,

$$\eta_{i'j'}dx^{i'}dx^{j'} + \frac{k(\eta_{m'n'}x^{m'}dx^{n'})^2}{1 - k\eta_{s't'}x^{s'}x^{t'}} = \eta_{ij}dx^{i}dx^{j} + \frac{k(\eta_{mn}x^{m}dx^{n})^2}{1 - k\eta_{st}x^{s}x^{t}}, \text{ or,}$$

$$d\vec{r}\,'\cdot d\vec{r}\,' + \frac{k(\vec{r}\,'\cdot d\vec{r}\,')^2}{1 - k\vec{r}\,'\cdot\vec{r}\,'} = d\vec{r}\cdot d\vec{r} + \frac{k(\vec{r}\cdot d\vec{r})^2}{1 - k\vec{r}\cdot\vec{r}}, \text{ or,}$$

$$\frac{d\vec{r}\,'\cdot d\vec{r}\,'(1 - kr'^2) + k(\vec{r}\,'\cdot d\vec{r}\,')^2}{1 - kr'^2} = \frac{d\vec{r}\cdot d\vec{r}(1 - kr^2) + k(\vec{r}\cdot d\vec{r})^2}{1 - kr^2}.$$

This can be proved with straightforward but lengthy algebra. Derive $\vec{r}\,'\cdot\vec{r}\,'$, $\vec{r}\,'\cdot d\vec{r}\,'$ and $d\vec{r}\,'\cdot d\vec{r}\,'$, in terms of $\vec{r}$, $d\vec{r}$ and $\vec{a}$, to begin the calculation.

2. A distant galaxy has a red shift $z = 0.25$. Using the red shift versus distance relation, what is the luminosity distance d_L and the proper distance now? What is the approximate proper velocity now, as given by the Hubble relation? What was the proper distance when light left the galaxy? Use the data for H_0 and q_0 in the text.

3. Show that $H(z) \approx H_0(1 - \frac{dH_0}{dt}z\frac{1}{H_0^2} + \cdots)$.

4. Using $\frac{dQ}{dt}/Q = H$, show $1 + z = \exp[\int_{t_1}^{t_o} H dt]$, where t_1 is the time light left the source at d, and t_o is the time observed at the origin. Also show that $H_1 = -\frac{dz}{dt}/(1+z)$. Note, many red shifts are observed at t_o so the red shift z is labeled by t_1, the time light was emitted.

5. Consider the conservation of energy and momentum $T^{\nu\mu}{}_{;\mu} = 0$ for a perfect fluid in a locally inertial frame. Show this leads to conservation of entropy $\frac{dS}{d\tau} = 0$ or adiabatic flow. Make use of Eqs. (9.23)–(9.27).

6. Assume the perfect fluid that represents the universe. Do not assume $k = 0$. Here $(\rho(t), p(t))$ are the (energy density, pressure) of the universe and are functions of time. Apply energy and momentum conservation and show that for the Robertson–Walker metric,

$$0 = \frac{d\rho}{dt} + 3(\rho + p)\frac{1}{Q}\frac{dQ}{dt} = \frac{d(\rho Q^3)}{dt} + p\frac{dQ^3}{dt}.$$

7. Use the results of Problem 6 to show that in the present universe $\rho \propto Q^{-3}$. Show that in the young radiation-dominated universe the equation of state was $p = \rho/3$.

8. In cosmology and astronomy, black-body radiation plays an important role. In this problem, the Planck spectrum is written in MKS units,

$$\frac{du}{d\lambda} = \frac{16\pi^2\hbar c}{\lambda^5}\frac{1}{exp[\frac{2\pi\hbar c}{\lambda k_B T}] - 1},$$

$$u = \frac{8\pi^5 k_B^4}{15(2\pi\hbar c)^3}T^4,$$

$$\lambda_{max}T = \frac{0.2014(2\pi)\hbar c}{k_B},$$

where du is the energy density in Jm^{-3} of the radiation in the region between wavelengths $(\lambda, \lambda + d\lambda)$ in m; T is the temperature in K; and $k_B = 1.381 \times 10^{-23}$ in JK^{-1} is Boltzmann's constant. The constants c and $\hbar$, are given in Chapter 1. The wavelength λ_{max} is that for which $\frac{du}{d\lambda}$ is a maximum. Evaluate the numerical values in the above equations in MKS units. Transform the equations to natural units and evaluate the numerical values. Note that T is the same in both sets of units.

9. Use the present values of the ordinary matter energy density, the radiation energy density and the temperature to calculate approximately

the number of baryons/photons. The proton rest energy is 938 MeV, and $k_B = 8.62 \times 10^{-11}$ MeV/K.

10. Hydrogen atoms could begin to form when the typical energy of a photon was about $0.015 E_I$, where E_I is the atom's ionization energy 13.6 eV. That's because there were too few higher energy photons in the Plank spectrum to ionize the atoms that were formed. From this time to the present, how much have the photons been red shifted? At that time, what was the ratio of the radiation energy density to the mass energy density, including dark matter? Use results from Problem 8, if needed.

11. In the text it was shown that for a radiation energy density-dominated universe $Q \propto t^{1/2}$. Now show that the typical photon energy is given by $k_B T = \alpha t^{-1/2}$ and evaluate α. Find the times in seconds when $k_B T = (1, 10^3, 10^6)$ MeV. Take $k = 0$ for the curvature constant. Hint: work in MKS units, use the Plank formula for the radiation energy density, express α in terms of $G, \hbar, c$ and then numerically evaluate α.

12. Using the data provided, calculate q_0, the present-day value of the acceleration parameter.

Bibliography

Adler, R., Bazin, M. and Schiffer, R. (1965). *Introduction to General Relativity* (McGraw-Hill).

Berlotti, B., Iess, L. and Tortora, P. (2003). *Nature* **425**, p. 374.

Chandrasekhar, S. (1983). *The Mathematical Theory of Black Holes* (Oxford University Press).

Clemence, G. M. (1947). *Rev. Mod. Phys.* **19**, p. 369.

COBE. http://lambda.gsfc.nasa.gov/product/cobe.

Drake, S. P. (2006). *Am. J. Phys.* **74**, p. 22.

Dyson, F. W., Eddington A. S. and Davidson C. (1920). *Phil. Trans. Roy. Soc. A* **220**, p. 291.

Einstein, A. (1905). http://einsteinpapers.press.princeton.edu/ **2** trans. 154, ibid., **6** trans. 129.

Event Horizon Telescope Collaboration (2019). *Astrophys. J. Lett.* **875**(L4), p. 1.

Everitt, C. F. W. *et al.* (2011). *Phys. Rev. Lett.* **106**, p. 221101.

Fixsen, D. J. (2009). *Astrophys. J.* **707**(2), p. 916.

Guth, A. (1977). *The Inflationary Universe* (Basic Books).

Hawking, S. W. (1975). *Comm. Math. Phys.* **43**, p. 199.

HIPPARCOS. http://esa.int.hipparcos.

Hubble, E. (1929). *Proc. Natl. Acad. Sci.* **15**, p. 168.

Hulse, R. A. and Taylor, J. H. (1982). *Astrophys. J.* **195**, p. L51.

Kaufman, W. J. (1985). *Universe* (W. H. Freeman and Company).

Kerr, R. P. (1967). *Phys. Rev. Lett.* **11**, p. 237.

Kruskal, M. D. (1960). *Phys. Rev.* **119**, p. 1743.

Lebach, D. (1995). *Phys. Rev. Lett.* **75**, p. 1439.

Lense, J. and Thirring, H. (1918). *Phys. Z.* **29**, p. 156.

LIGO. http://ligo.org.

Möller, C. (1952). *The Theory of Relativity* (Oxford University Press).

Perlmutter, S. *et al.* (1999). *Astrophys. J.* **517**, p. 565.

Perrin, R. (1979). *Amer. J. Phys.* **47**, p. 317.

Peters, P. C. (1964). *Phys. Rev. B* **136**, p. 1224.

PLANK (2015). http://lanl.arXiv.org/astro-ph/arXiv:1502-01582.

Pound, R. V. and Rebka, G. A., Jr. (1960). *Phys. Rev. Lett.* **3**, p. 554. ibid., **4**, p. 337.

Riess, A. *et al.* (1998). *Astron. J.* **116**, p. 1009.

Robertson, H. P. (1935). *Astrophys. J.* **82**, p. 284, ibid., **83**(1936), p. 187, 257.

Rubin, V., Ford, V. and Kent, W. Jr. (1970). *Astrophys. J.* **159**, p. 1379.

Schiff, L. (1960). *Phys. Rev. Lett.* **4**, p. 215.

Schutz, B. (2009). *A First Course in General Relativity* (Cambridge University Press).

Schwarzschild, K. (1916). http://arxiv.org/abs/physics/9905030.

Shapiro, I. I. (1964). *Phys. Rev. Lett.* **13**, p. 789.

Shapiro, I. I. *et al.* (1971). *Phys. Rev. Lett.* **26**, p. 1132.

Soszynski, I. *et al.* (2012). *Acta Astron.* **62**, p. 219.

Suzuki, H. *et al.* (2019). *Mon. Nat. R. Astron. Soc.*, p. 1.

Thorne, K. (2014). *The Science of Interstellar* (W. W. Norton).

Walker, A. G. (1936). *Proc. London Math. Soc.* (2) **42**, p. 90.

Weinberg, S. (1972). *Gravitation and Cosmology, Principles and Applications of the General Theory of Relativity* (John Wiley & Sons).

Weinberg, S. (1977). *The First Three Minutes* (Basic Books).

Weisberg, J. M., Nice, D. J. and Taylor, J. H. (2010). *Astrophys. J.* **722**, p. 1030.

Will, C. M. (2006) *Living Rev. Relativity* **9**, p. 3.

WMAP. http://map.gsfc.nasa.gov/product/cobe.

Index